Cell Diagnostics
Images, Biophysical and Biochemical Processes in Allelopathy

T0174383

Cell Diagnostics

Images, Biophysical and Biochemical Processes in Allelopathy

Editors

V.V. Roshchina

Laboratory of Microspectral Analysis of Cells and Cellular Systems
Russian Academy of Sciences
Institute of Cell Biophysics
Institutskaya Street 3, Pushchino, Moscow region, 142290, Russia.
E-mail: roshchina@icb.psn.run

S.S. Narwal

Department of Agronomy,
CCS Haryana Agricultural University,
Hisar-125 004, India
E-mail: narwal_1947@yahoo.com

CRC Press
Taylor & Francis Group
Boca Raton   London   New York

CRC Press is an imprint of the
Taylor & Francis Group, an **informa** business

A SCIENCE PUBLISHERS BOOK

First published 2007 by Science Publishers Inc.

Published 2019 by CRC Press
Taylor & Francis Group
6000 Broken Sound Parkway NW, Suite 300
Boca Raton, FL 33487-2742

First issued in paperback 2019

No claim to original U.S. Government works

ISBN 13: 978-0-367-44622-2 (pbk)
ISBN 13: 978-1-57808-510-1 (hbk)

**Visit the Taylor & Francis Web site at
http://www.taylorandfrancis.com**

**and the CRC Press Web site at
http://www.crcpress.com**

Library of Congress Cataloging-in-Publication Data

Roshchina, V.V. (Viktoriia Vladimirovna)
  Cell diagnostics: images, biophysical and biochemical processes in allelopath/V.V. Roshchina, S.S. Narwal.
     p. cm.
  Includes bibliographical references and index.
  ISBN 978-1-57808-510-1 (hardcover)
  1. Allelopathic agents--Research--Technique. 2. Allelopath--Research--Technique. I. Narwal, Shamsher S. II. Title.

QK898.A43R67 2007
571.9'2--dc22

                                                          2007019672

Foreword

Allelopathic interactions arise from the production of secondary metabolites by plants and microorganisms, that produce a wide array of biochemical compounds that create biological changes either in the organism that produces it or in adjacent organisms. It is viewed as a causal factor in increasing the yield of agriculture and forestry. These important changes are being studied world over in cellular and allelopathy laboratories and field experiments. They are being incorporated into the production of modern crops and modern forestry. The environmental "well being" of countries where allelopathy has been recognized, has produced a better standard of living.

Allelopathy occurs in most natural communities and agroecosystems, but frequently it is unrecognized. The adverse effects from allelochemicals arising from certain weeds and crops reduce production in agricultural fields and managed forest systems. There is a need to evaluate the allelopathy effects of previous plants, residues, associated plants, that may occur in cropping systems. It is possible to manage allelopathy in weed control in agriculture and forestry. This is being accomplished in some countries through management and in some situations the authorities rely heavily on analytical chemistry. Management is important since it involves residues manipulation, desirable planting of cover crops, crop rotation, fertilizer and water practices and adjusting herbicide application to complement allelopathic weed control. Sunflower, grain sorghum and rye are effective weed control crops. Evidences show that allelochemicals and herbicides have an added negative impact on sensitive species, such as wheat, cotton, and corn. Management practices must be developed that focus on allelopathy and herbicides to arrive at sustainable agricultural practices that have less deleterious effects on the environment.

Allelopathy arises because growth stimulating, or inhibiting plant and microbial produced biochemicals which are released into the environment. The stress conditions such as moisture, temperature, fertilizer, soil, pests, biotic factors. etc., that alter the growth of plants and microorganisms produce more allelochemicals. Because it is almost impossible to separate these factors in field situations, much of our understanding of allelopathy comes from bioassays that include cellular microscopy and analytical chemistry techniques.

This book outlines some fields of cell diagnostics applied to allelopathy with several chapters devoted to Cellular Model Systems for Allelopathy, New Methods of Microscopy in Cellular Diagnostics and Methods of Analytical Biochemistry and Biophysics. It may be useful both for the study of allelopathic mechanisms at cellular level and for the tests for the allelochemicals' screening. Cellular level of the investigation in the chemical relations between living organisms is represented as applications to study the allelopathic events with modifications of electron microscopy, luminescent microscopy, laser scanning confocal microscopy, optical coherence microscopy and microspectrofluorimetry as well as analytical biochemical and biophysical assays. This book written by a distinguished group of international authors. Although the book does not give a complete account of these fields, we hope that it will serve as the starting point to help guide the research of others.

Multidisciplinary approaches will be required before the tools of biotechnology can be applied to the production of agriculture and forestry plants that produce allelochemicals that aid in their own pest and weed control. There is much optimism and little progress made toward unraveling the complexities of analytical chemistry, physiological, biochemical and biotechnological interactions between species. We need a firmer foundation describing the existence and functions of allelopathy. We wish to maximize the opportunities presented by allelopathy in our world today and explain its multiple roles in ecology, agriculture and forestry.

George Waller
Professor Emeritus
Department of Biochemistry and Molecular Biology
Oklahoma State University
Room 246, Noble Research Center
Stillwater, OK 74078
Foundation President
International Allelopathy Society
E-mail: gwaller@biochem.okstate.edu

Preface

Allelopathy is newly emerging multidisciplinary field of agricultural research. A lot of allelopathy research work has been done in various fields of agriculture and plant sciences. However, standard methods are not being used by workers due to lack of a compendium on the techniques, and hence the results obtained are not easily comparable with each other. This causes problems to researchers working in underdeveloped/third world countries in small towns, where library and research facilities are not available. Therefore, to make available the standard methods for conducting allelopathy research work, this multi-volume book has been planned, with one volume each for each discipline. In all the conferences held since 1990's a need has always been felt for a manual on allelopathy research methods. This book series aims to provide basic information about various methods to research workers, so that they can conduct research independently without the requirement of sophisticated equipments. The methods have been described in a simple way just like a DO IT YOURSELF book.

In allelopathy studies, the allelochemicals first influence the physiological and biochemical processes in cells. Till now there is no book of methods to study allelopathic interactions in the cells. The activity of cells influence various important physiological processes like seed germination, plant growth and development, photosynthesis and respiration, senescence and abscission are included in this volume. To understand the basic mechanisms of various physiological processes, being affected by allelochemicals at the cellular level enzyme activity and metabolite studies are essential.

The book is divided into 3 sections. Section I.Cellular Model Systems, includes 3 chapters (Allelopathy and plant cell diagnostics, Cellular models as biosensors and microalgae for determining the effects of allelochemicals). Section II. New Methods of Microscopy has 5 chapters (Microscopic methods to study morpho-cytological events during the seed germination, optical coherence tomography and optical coherence microscopy to monitor water absorption, optical coherence microscopy : study of plant secretory structures, laser-scanning confocal microscopy (LSCM) : Study of plant secretory cells and Luminescent cell analysis in allelopathy). Section III. Methods of Analytical Biochemistry and Biophysics consists of 6 chapters (Biochemical approach to study oxidative damage in plants exposed to allelochemical stress: A case study, Cholinesterase activity as biosensor reaction for natural allelochemicals: Pesticides and pharmaceuticals, Effects of allelochemicals on Algae membrane integrity, Total phenolics and phenolic acids in plants and soils, Computational methods to study properties of allelochemicals and modelling of molecular interactions in allelopathy and allelopathic pollens: isolating the allelopathic effects).

This book will serve as ready reference in the laboratory or class room and help to solve many problems of cell studies in agriculture and allied fields including allelopathy. Information provided can be use to determine the effects and mechanism of action of allelochemicals at the cellular levels. It will be useful for undergraduate and graduate students pursuing allelopathic work, plant physiologists, biochemists and other plant science specialists. We have tried to provide appropriate solutions to the problems of cell studies. The users of this book can select suitable methods, according to the available facilities.

We are indebted to the contributors, who have actually used all these methods in their fields of specialization for the last 10 to 25 years, in accepting challenging task to prepare chapters for the present book. All of them have made sincere efforts in presenting the procedure of various methods in very simple language, easily understood by beginners.

We will appreciate to receive valuable suggestions from the students and researchers, to make further improvements in future editions of this book, to make it more useful and meaningful.

December 3, 2006   **V.V. Roshchina**
S. S. Narwal

Contents

Detail of Contents

List of Contributors

Aliotta Giovanni, Professor of Botany. Dipartimento di Scienze della Vita, Seconda Università di Napoli, Via Vivaldi 43, I - 81100 Caserta, Italy. Dipartimento di Scienze della Vita, Seconda Università di Napoli, Via, E-mail: aliotta@unina.it, Fax:++390823274571, Tel.0823274561, Fax: 0823 274571

Anaya, Ana Luisa, Professor, Departamento de Ecología Funcional, Instituto de Ecología, Universidad Nacional Autónoma de México, Circuito Exterior, Ciudad Universitaria, 04510 México, D.F., Phone (52-55) 5622-9032, Fax (52-55) 5622-9043, E-mail: alanaya@miranda. ecologia.unam.mx; rcruz@miranda.ecologia.unam.mx, alanaya@ miranda.ecologia.unam.mx; Mexico

Budantsev, Arkady Yustianovitch, Professor, Head of Department of Analytical Microscopy, Russian Academy of Sciences, Institute of Theoretical and Experimental Biophysics, Istitutskaya Str., 3, Pushchino, Moscow Region, 142290, Russia, E-mail: budantsev@mail. ru, Fax: (7-0967)790553

Cafiero Gennaro, Director. Centro Interdipartimentale di Servizio per la Microscopia Electronica Università di Napoli Federico II, Via Foria 223, I-80139 Napoli, Italy, E-mail: gecafier@unina.it, Fax: ++ 39 081 440507

Ciniglia, Claudia, PhD, researcher, Dipartimento di Scienze della Vita, Seconda Università di Napoli, ViaVivaldi 43, I- 81100 Caserta, Italy. Claudia Ciniglia,. E-mail: ciniglia@unina.it., Fax: ++390823 274571

Cruz-Ortega, R. Mexico, Ph.D. researcher, Departamento de Ecologia Funcional, Instituto de Ecología, Universidad Nacional Autónoma de Mexico, Circuito Exterior, Ciudad Universitaria, 04510 México, D.F., Phone (52-55) 5622-9032 Fax: (52-55) 5622-9043, E-mails: arcruz@miranda.ecologia.unam.mx

Djurdjevic Lola, Professor .Dr . Sci., Department of Ecology, Institute for Biological Research "Sinisa Stankovic"m Bulevar Despota Stefana 142 11060 Belgrade, Serbia and Montenegro, Phone: +381 11 20 78 359, Fax: +381 11 27 61 433, E-mail: kalac@ibiss.bg.ac.yu

Fergola Paolo, Professor of Mathematic, Dipartimento di Matematica e applicazioni "Renato Caccioppoli", Università, di Napoli Federico II, Via Cinthya, Napoli, Italy, E-mail paolo.fergola@unina.it

Gelikonov Grigory V, Dr.Sci., Professor, Institute of Applied Physics RAS, Uljanova st., 46, 603600 Nizhny Novgorod tel., 7(8312)368010, E-mail: vlad@ufp.appl.sci-nnov.ru

Giordano Simonetta Professor, Università degli Studi di Napoli "Federico II", Dipartimento di Biologia Strutturale e Funzionale, Complesso Universitario Monte S.Angelo, Via Cinthia, 80126 Napoli, Italy. E-mail: simonetta.giordano@fastwebnet.it. Present address:Via Foria, 223, I - 80139 NAPOLI, +39 081 2538553 (studio), +39 081 2538556 (laboratorio), Fax +39 081 2538523. E-mail: giordano@unina.it, http://www.docenti.unina.it/simonetta.giordano

Hong-Ying Hu. Professor, Environmental Simulation and Pollution Control State Key Joint Laboratory, Department of Environmental Science and Engineering, Tsinghua University, Beijing 100084, China. E-mail: lfm01@mails.tsinghua.edu.cn

Kamensky Vladislav A, Ph.D.Nizhny Novgorod State University, Department of biology, Gagarina 23, 603091 Nizhny Novgorod, Russia. E-mail: vlad@ufp.appl.sci-nnov.ru

Kononov, Alexey. V., Magister of biophysics, system administrator, Russian Academy of Sciences Institute of Cell Biophysics, Institutskaya Str., 3, Pushchino, Moscow Region, 142290, Russia. Present address: Kolkhoznaya str., 8, Pilna, Nizhny Novgorod Region, 607490, Russia. E-mail: pilna@mail.ru

Kuranov Roman L. Dr.Sci, Nizhny Novgorod State University, department of biology, Gagarina 23, 603091 Nizhny Novgorod, Russia. E-mail: vlad@ufp.appl.sci-nnov.ru

Kutis Irina S., researcher, PH.D. Institute of Applied Physics RAS, Uljanova st., 46, 603600 Nizhny Novgorod, Russia tel.: 7(8312)368010, E-mail: vlad@ufp.appl.sci-nnov.ru

Kutis Lev S., Dr.Sci., Professor, Nizhny Novgorod State University, department of biology, Gagarina 23, 603091 Nizhny Novgorod, Russia. E-mail: vlad@ufp.appl.sci-nnov.ru

Kutis Sergey D, researcher, Ph.D. Institute of Applied Physics RAS, Uljanova st., 46, 603600 Nizhny Novgorodtel. +7(8312)368010, E-mail: vlad@ufp.appl.sci-nnov.ru

Li Feng-Min, reseacher, Ph.D. Environmental Simulation and Pollution Control State Key Joint Laboratory, Department of Environmental Science and Engineering, Tsinghua University, Beijing 100084, China. E-mail: lfm01@mails.tsinghua.edu.cn

Lo Piparo Elena, Professor, Bioinformatics Group, Department of BioAnalytical Science, Nestle Research Center P.O. Box 44, CH-10000 Lausanne 26, Switzerland Tel* : +41 (0) 21 785 9530 * : +41 (0) 21 785 9486. E-mail: Elena.LoPiparo@rdls.nestle.com

Murphy Stephen D., ES1 209 Dept of Environment and Resource Studies, 200 University Avenue West, Waterloo ON N2V 2L1 Canada, vox 519 888 4567 x5616, fax 519 746 0292, E-mail: sd2murph@fes.uwaterloo.ca Vice-Chair, Society for Ecological Restoration Ontario, Co-Director Research, Cruickston Charitable Research Reserve.

Pavlovich Pavle. Ph.D., Department of Ecology, Institute for Biological Research "Sinisa Stankovic"m Bulevar Despota Stefana 142 11060 Belgrade, Serbia and Montenegro, Phone: +381 11 20 78 359, Fax: +381 11 27 61 433, E-mail: kalac@ibiss.bg.ac.yu

Petriccione Milena, PhD student , Dipartimento di Scienze della Vita, Seconda Università di Napoli, Via, Vivaldi 43, I- 81100 Caserta, Italy.

Pinto Gabriele, Associate Professor of Botany. Dipartimento delle Scienze Biologiche, Sezione Biologia Vegetale, Università di Napoli Federico II, Via Foria 223, I-80139 Napoli, Italy. E-mail: gabpinto@unina.it. Fax: ++39081 450165

Pollio Antonino, Associate Professor of Botany. Dipartimento delle Scienze Biologiche, Sezione Biologia Vegetale, Università di Napoli Federico II, Via Foria 223, I-80139 Napoli, Italy. E-mail: lanpollio@unina.it. Fax: ++39081 450165

Roshchina Victoria V., Dr.Sci., Professor, Institute of Cell Biophysics RAS, Institutskaya Str., 3, Pushchino, Moscow Region, 142290, Russia. E-mail: roshchina@icb.psn.ru, Fax : 7 (0967) 330509; Tel: 7 (495)9237467

Sapozhnikova Veronika V, Dr.Sci , Institute of Applied Physics of RAS, Uljanov st. 46 Nizhny Novgorod, Russia, 603950, Fax: +7 (8312) 363792, Phone: +7 (8312) 3680-10; E-mail: sapozhnikova@ufp.appl.sci-nnov.ru or vsap1972@mail.ru new address: University of Texas of Medical Branch (UTMB), Galveston, 77550, USA

Shabanov Dmitry V. Ph.D. Institute of Applied Physics of RAS, Nizhny Novgorod, 603950, Russia, E-mail: vlad@ufp.appl.sci-nnov.ru

Waller, George. R. Professor Emeritus, Department of Biochemistry and Molecular Biology, Oklahoma State University, Room 246, Noble Research Center, Stillwater, OK 74078, USA. Tel: 405-744-6692, Fax: 405-744-7799, E-mail: gwaller@biochem.okstate.edu

Yashin, Valerii A., Head of Optic Department, Russian Academy of Sciences Institute of Cell Biophysics, Institutskaya Str., 3, Pushchino, Moscow Region, 142290, Russia. E-mail: Yashin@psn.ru

Yashina, Alexandra V. Young Scientist, Engeneer, Russian Academy of Sciences Institute of Cell Biophysics, Institutskaya Str., 3, Pushchino, Moscow Region, 142290, Russia. E-mail: Yashina@psn.ru

Section 1

Cellular Model Systems

Chapter

Allelopathy and Plant Cell Diagnostics

S.S. Narwal and V.V. Roshchina

1. INTRODUCTION

Allelopathy refers to any process involving secondary metabolites produced by plants, microorganisms, viruses and fungi that influence the growth and development of agricultural and biological systems. It has been established that allelopathy offers great potential (a) to increase agricultural production (food grains, vegetables, fruits, forestry), (b) to decrease harmful effects of modern agricultural practices [multiple cropping, leaching losses from N fertilizers, indiscriminate use of pesticides (weedicides, fungicides, insecticides, nematicides), tolerant/ resistant biotypes in pests] on soil health/productivity and on environment and (c) to maintain soil productivity and a pollution- free environment for our future generations. It is likely that in the near future allelopathy will be used in crop production, crop protection, agroforestry and agrohorticultural practices in developed and developing countries. Allelopathy may become one of the strategic sciences to reduce environmental pollution and to increase agricultural production in sustainable agriculture in the 21st Century. Allelopathy provides a basis for sustainable agriculture, hence, currently allelopathic research is being carried out in most countries worldwide and is now receiving more attention from agricultural and bioscientists.

The term 'allelopathy' was coined by Prof. Hans Molisch, a German plant physiologist in 1937. It is a new field of science and, till now there is no Book **on Methodology of Allelopathy Research.** Thus causing a lot of problems to researchers working in underdeveloped/Third World countries, in small

Department of Agronomy, CCS Haryana Agricultural University, Hisar-125 004, India
E-mail: narwal_1947@yahoo.com

towns without library facilities. Therefore, this multi-volume book has been planned so as to make available the standard methods for conducting allelopathic research independently. Since allelopathy is a multi-disciplinary area of research, hence, volumes have been planned for each discipline.

2. CELLS RESPONSES TO ALLELOCHEMICALS

The allelochemicals are secondary substances, biosynthesized from the metabolism of carbohydrates, fats and amino acids and arise from acetate or the shikimic acid pathway. These are biosynthesized and stored in the plant cells and do not affect the cell activities. However, after their release from the plant cells (through volatilization, leaching root exudates and decomposition of biomass), these allelochemicals start influencing the organisms (plants, pathogens, insect pests etc), when they come in contact with them. Rice (1984) indicated that allelochemicals agents influence the plant growth through following physiological processes : (a) cell division and cell elongation, (b) phytohormone induced growth, (c) membrane permeability, (d) mineral uptake, (e) availability of soil phosphorus and potash, (f) stomatal opening and photosynthesis, (g) respiration, (h) protein synthesis and changes in lipid and organic acid metabolism, (i) inhibition of porphyrin synthesis, (j) inhibition or stimulation of specific enzymes, (k) corking and clogging of xylem elements, stem conductance of water and internal water relations and (l) miscellaneous mechanisms. Most of these physiological and biochemical processes are responses of cells to various allelochemicals.

3. PLANT CELL DIAGNOSTICS

The plant cell is a basic unit for germination, growth and development of plants. The allelochemicals first come in contact with the cell and then allelopathic interactions take place. Some allelochemicals have broad-spectrum activity that extends to the tissues of host plants, where their effects may be either beneficial or deleterious to plant germination, growth, development or yield. Hence, to understand the mechanisms of such intercations interactions the study of the cell and its various processes is very necessary. Therefore, this book has been been prepared (a) to make available all methods for such studies and (b) scientists can understand the scope of allelopathic research in relation to the cell. Hence, we have explained and discussed various techniques to study cell processes etc.

4. REFERENCES

Narwal, S. S. (1994). *Allelopathy in Crop Production*. Scientific Publishers, Jodhpur, India.
Rice E. L. (1984). *Allelopathy*. II edition. Academic Press, New York, USA.

Chapter

Cellular Models as Biosensors

V.V. Roshchina

1. INTRODUCTION

Biosensors are the analytical systems, which contain sensitive biological elements and detectors. Plant cells as a possible biosensors have natural structure that determinates their high activity and stability. Criteria in the screening of the plant cells as biosensors for allelopathy should be as under: (i) Reaction is fast based on the time of response, (ii) Reaction is sensitive to small doses of analysed compounds or their mixtures and (iii) Methods of detection viz., biochemical, histochemical, biophysical (in particular, spectral changes in absorbance or fluorescence) are easy in laboratory and in the field conditions. The search of biosensors in active plant species is suitable to determine the mechanisms of action of biologically active substances or external factors of the environment (Roshchina and Roshchina, 2003; Roshchina, 2004; 2005 c)).

The search for natural biosensors is aimed at cells with most sensitive reactions to biologically active compounds. Such reactions are important because they show changes in the process of recognition during the chemical interactions of plant cells, including pollen-pistil contacts during the fertilization of seed-bearing plants (Roshchina, 1999; 2001a,b), germination of vegetative microspores in Cryptogams (Roshchina, 2004; 2005 a,b) or interaction of plant cells and insect as in carnivorous plant species (Muravnik and Ivanova, 2002). Test-objects may be cells, which

Laboratory of Microspectral Analysis of Cells and Cellular Systems, Russian Academy of Sciences Institute of Cell Biophysics, Institutskaya str., 3, Pushchino, Moscow region, 142290, Russia. E-mail: roshchina@icb.psn.run serve in sexual [pollen (male gametophyte) and pistil (female gametophyte) in flower] and vegetative breeding (vegetative microspores), respectively. In this chapter, the methods to determine the reactions of such cells are described.

2. CELLULAR BIOSENSORS AND THEIR REACTIONS

The biosensors for allelochemicals are: (i) vegetative microspores of spore-bearing plants, (ii) pollen as generative (male) microspores of seed-bearing plants (iii) pistil stigma (female gametophyte) of seed-bearing plants and (iv) secretory cells of carnivorous plants (Roshchina, 1999; 2001a,b; 2004). The germination and fluorescence of plant microspores, both in generative (pollen or male gametophyte of Golosperms and Angiosperms) and in vegetative (in Cryptogams), may be used as test-reactions after the addition of allelochemical. The microspores serve for sexual and vegetative breeding and may germinate on artificial media. Some enzymic reactions (e.g. the cholinesterase activity) are also tested in the microspores.

2.1   Vegetative Microspores of Spore-Producing Plants

Vegetative microspores of Cryptogams are unicellular objects with hard cover, blue-fluorescence in ultra-violet light and elaters [serve as anchor to a substrate (soil)]. The cells are diploid and have autotrophic nutrition due to the presence of chloroplasts, where photosynthesis occurs. They germinate well in artificial nutrient medium or in water and ultra-violet light induces significant autofluorescence.

Principle: Vegetative microspores of spore-bearing (Cryptogams) plants, particularly of horsetail [*Equisetum arvense L.* (Fig.1)] are unicellular objects suitable for use, due to their size for the microscopic observations in cell biology. The cells are visible even at minimal objective x 10 of common transmission and luminescent microscopes. Advatages of the semicrospores as object are: (i) Faster division after moistening for 1-2 h not more than 24 h, (ii) germinate well in artificial nutrient medium or in water, (iii) changes in cell interior and cell wall are visible under usual and luminescent microscope, (iv) sensitivity to natural substances and (v) significant autofluorescence recorderd after 1-2 min or 1-2 h of moistening in control experiments.

Materials required: Vegetative microspores of *Equisetum arvense* L. [collected from the meadows in end April or in May and stored in dry conditions] ordinary and luminescent microscopes with photocamera (one microscope with luminescent and transparent regimes), and if possible, microspectrofluorimeters (see Chapter 9) or Laser Scanning Confocal Microscope (see Chapter 8), object glasses, Petri dishes, Potassium phosphate, Calcium chloride, Sodium chloride, Magnesium chloride, Kodak colour–400 photofilm.

Procedure: Vegetative microspores of horsetail (*Equisetum arvense*) fluoresce and germinate on the object glasses (slides) after the moistening with artificial medium or simple water (Roshchina *et al.*, 2002;2003; 2004). The nutrient medium contains: 6.63 µg/L K phosphate, 6.51 µg/L Cà chloride, 3.47µg/L Na chloride, 5µg/L Mg chloride 5. Tested compounds and extracts were also dissolved or diluted in above- mentioned nuitrient medium. All experiments were performed at room temperature (20-22 °C). The growth occurred in the solution with allelochemical or other tested substances [(0.05 ml = 1 drop)] on the object glasses kept on the wet paper in Petri dishes. When the action of volatile compounds is studied, the small cap (0.5-1 cm diameter) with the sample is kept on the same Petri dish near the object glass with microspores moistened by the nutrition medium. Five ml of water was added to the bottom of every dish and 4-5 dishes with the slides were used per treatment. The volume of one drop was 0.05 ml for the microspore to develop one slide. The number of germinated vegetative microspores (red-fluorescing) was counted using objective x 10, 20 or 40 under luminescent microscope. When vegetative microspores from *Equisetum arvense* develop, new molecules of chlorophyll are formed, which fluoresce in the red spectral region (Fluorescing vegetative microspores

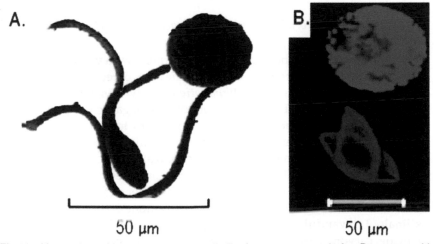

Fig. 1   Vegetative microspores of horsetail *Equisetum arvense*. Left - Dry spore with elaters fluoresce under UV light (360-380 nm); Right - Germinated spore without elaters. The blue-fluorescing cover is missed.

could be photographed, using Kodak colour –400 photofilm). Hence, the blue fluorescence of non-germinated spores (when the rigid cover is on the microspore cell) change to red fluorescence of germinated ones (after the cover has been removed, the cell starts to divide and the content of chloroplasts is increased), 2-24 h after moistening (Fig.1). The number of germinated vegetative microspores (red-fluorescing) was counted using the luminescent microscope. The changes in the germination rates of vegetative microspores of *Equisetum arvense* are considered as possible biosensor reactions to study the allelopathic mechanisms.

Observations: Observation of each test sample in vegetative microspores of *Equisetum arvense* lasts from 2 h (for fresh microspores, < 1 month after collection) to 6-24 h (for microspores stored > 1 month). Observer fixes the start of development as a missing of rigid blue-fluorescing cover and elaters, and after 2-24 h may count the number of red-fluorescing microspores that increase in the red chlorophyll fluorescence in comparison to undeveloped (fluoresce in blue) and see the divided cells. Under ordinary microscope, after 2-3 days each microspore may form rhizoid cells, which have less amount of chloroplasts than prothallium (first divided cells) and thallus cells giving gametophyte.

Statistical analysis: One hundred microspores were analysed per slide. Counting was done in four or five replicates (the number of Petri dishes per treatment). Results were expressed as mean ± SEM (SEM shown graphically in figures). The relative standard deviation (RSD) was 5-6% (n = 400-500 microspores per one variant; P =0.95).

Experiment 1. *Effects of allelochemicals on the germination of horsetail vegetative microspores.*

In Table 1 the results of the microspores' treatment with volatile excretions from lavender (*Lavandula* sp.)oil, disrupted the fruits of red pepper (*Capsicum annuum*) and cut bulb of garlic (*Allium sativum*) as well as the water extracts from same samples. Lavender oil or cut bulb of garlic were put in Petri dish, in small vessels near the object glass with test microspores. The volatiles produced from the materials exert influence on them via air. Besides, water (nutrient medium for microspores) extracts from fruits of red pepper and the garlic bulbs (1: 1000 w/v) were prepared and added directly to the microspores on the object glasses. Test lavender oil from fresh lavender flowers contained linalyl acetate (up to 40 %) among 100 components such as linalool, lavandulol, lavandulyl acetate, terpineol, cineol, limonene, ocimene, caryophyllene, etc (Roshchina and Roshchina, 1993; Golovkin *et al.*, 2001). The fruits of red pepper release biologically active alkaloid capsaicin, which may be in volatile and soluble form. Garlic volatile excretions contain thiocyanates alliin and allicin, whereas, their water extracts from bulbs have diallyl disulphide (Roshchina and Roshchina, 1993).

Table 1 The effects of volatile plant excretions and water extracts (1: 1000) w/v) from allelopathic species with pesticidal features on the germination of horsetail vegetative microspores.

Excretion	Microspores Germination (% of control)		
	Near slide	2.5 cm distance	Water extract from the same probe (1:1000)
Lavender (*Lavandula vera*) oil	64.5± 6	98.5 ± 6	–
Disrupted fruits (79 mg) of *Capsicum annuum*	139± 9	125± 10	88 ± 14
Cutted bulb of garlic *Allium sativum* L. (350 mg)	154± 10	132± 10	104± 6

The effects on germination of microspores depend on the concentration and the distance from the object glass with microspores moistened with nutrient medium lavender oil (active matter) inhibit or not the process, while the volatile exretions from other plant species have no inhibitory, but stimulatory effects. Only water extract from red pepper demonstrates weak activity.

Experiment 2. *Effects of liquid allelochemicals on the germination of vegetative microspores.*
Results of experiments with water and ethanol extracts from sesquiterpene-enriched plants *Achillea millefolium* and *Gaillardia pulchella* are given in Table 2. The inhibitory effects are dependent on the solvents and the concentration of plant material. Strongest effects were observed mainly, for leaves and flowers.

Experiment 3. *Effects of individual compounds on the germination of vegetative microspores.*
Colchicine is allelochemical from *Colchicium* genera, which binds the tubulin and prevents the mitosis (Fig. 2) Cytochalasin B, cell permeable fungal toxin from *Helminthosporium dematiodeum*, which inhibits cell division by blocking the active polymerization and formation of contractile actomyosin microfilaments, inhibit the germination of microspores (Roshchina, 2005a). Therefore, one mechanism of action of some allelochemicals from plants and

Table 2   The effects of water and ethanol extracts from allelopathically active plant species on the germination of *Equisetum arvense* microspores (amount of red-fluorescing microspores, % of control)

Plant species	Organ	Water extract (1:10 w/v)	The ethanol extracts * 1: 1000	1:100
Achillea millefolium	Flowers	83±20	161±22	78±17
	Leaves	39±10	117±20	82±20
	Roots	121±13	117±20	109±21
Gaillardia pulchella	Flowers	143±20	87±17	26±10
	Leaves	82±21	9±2	9±3
	Roots	191±30	96±17	17±4

Original ethanolic extracts (1:10 w/v), were mixed with water to get final concentration. The ethanol concentration was not more 1/100 or 1/1000 and had no effect on the studied reaction.

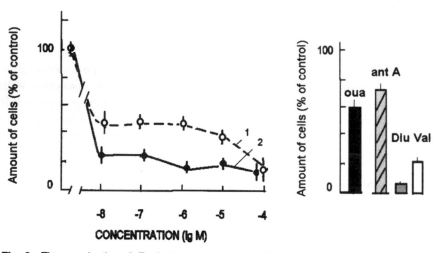

Fig. 2 The germination of *Equisetum arvense* vegetative microspores in the presence of anticontractile agents (left) and energetic inhibitors 10^{-5}M (right). 1- colchicine; 2- cytochalasin B; oua – ouabain, ant – antimycin A; diu – diuron, val – valinomycin.

fungi that depress the plant growth occurs through the inhibition of cellular motility. The other mechanisms of retardation in microspore germination are the effects of allelochemicals as energy inhibitors from plants and fungi (Fig. 2). For example, similar modes of action are peculiar to ouabain (from *Strophanthus* genera, selective inhibitor $Na^+ - K^+$ ATPase), antimycin A (from *Streptomyces* sp., inhibitor of respiration in all organisms and cyclic electron transfer in autotrophs' photosynthesis), diuron (artificial inhibitor of non-cyclic electron transport in photosynthesis and valinomycin) and K^+-selective ionophoric cyclodepsipeptide.

2.2 POLLENS OF SEED-BEARING PLANTS

Generative microspores are pollen or male gametophyte of seed-bearing plants (Golosperms and Angiosperms). They are unicellular objects with hard cover that contains polymers of phenols or carotenoids. Unlike vegetative microspores, pollen grains produce male gametes-sperm with haploid genome. Pollens without elaters, cannot develop as autotrophic cells and lacks chloroplasts.

Principle: Pollen development may be observed either on the pistil stigma (*in-vivo*) or on the artificial medium (*in-vitro*). After the moistening (by secretion of flower pistil or artificial medium), this vegetative cell forms the male gametes called spermia (which may move with forming pollen tube to the tip). When pollen is developed on a flower pistil, spermia moves to the ovule. Then one spermia copulated with the egg cell, both nuclei are combined and the fertilization occurs. Pollen from knight star' [*Hippeastrum hybridum* L. (Herb.)], clivia (*Clivia hybrida*) and mock-orange (*Philadelphus grandiflora*) forming pollen tube, were visible even at minimal objective x 10 of the microscope. Nucleus in pollen, spermia and pollen tubes could be observed, respectively in light and luminescent microscopes, by using objective x 10, 20 and 40 (Fig. 3). The cells were photographed on highly-sensitive photofilm used for aeroshooting from aeroplane. Pollen

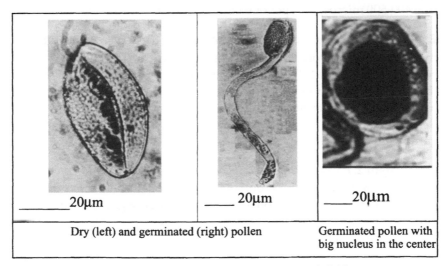

_____20μm	_____ 20μm	_____20μm
Dry (left) and germinated (right) pollen	Germinated pollen with big nucleus in the center

Fig. 3 Pollen from knight star' *Hippeastrum hybridum*

allelopathy plays important role in the normal plant fertilization in biocenosis (Murphy, 1999; Roshchina, 1999; Roshchina and Melnikova, 1999).

Materials required: Pollens were collected from the green house or in nature, light microscope, object glasses, Petri dishes.

Procedure: Pollen develops on the nutrition medium, forming pollen tube (Fig. 3). The nutrient medium for pollen was 10% sucrose solution and tested compounds were also dissolved in 10% sucrose. The pollen develops till the formation of pollen tube that lasts from 2-3 h (fresh collected and one week stored pollen) to 24 h (stored > 1 week, but < 1.5-2 months). All experiments were done at room temperature 20-22 ^{O}C. The growth occurred in the solution studied (0.05 ml = 1 drop) on the slides (object glasses) put on wet paper in Petri dishes. Five ml of water was added to the bottom of every dish and 4-5 dishes with the slides were used per treatment. Using light microscope, we determined the microspores germination (%) 2-24 h after moistening. The number of developed pollen tubes was counted.

Observations: Observation on each test sample on pollen lasts from 2 –24 h (fresh microspores, < 1 month after collection) to 6-24 h (microspores stored > 1 month).

Statistical analysis: One hundred pollen grains were examined on each slide and 100 microspores were analysed per slide. Counting was done in four or five replicates (the number of Petri dishes per treatment). Results were expressed as mean ± SEM. The relative standard deviation (RSD) was 5-6% (n = 400-500 microspores per one variant; P =0.95).

Experiment 1. *Effects of volatile allelochemicals on development of pollen tubes.*
In this experiment, volatile and liquid excretions from plants with pesticidic properties were tested (Table 3). The development of pollen tubes depend on the concentration and the distance from the object glass with microspores moistened with nutrient medium vapors of lavender oil (active matter) depress the process as well as red pepper, but garlic not. Water extracts of garlic were more effective.

Experiment 2. *Effects of terpenoid allelochemicals on pollen germination*
The effects of the terpenoids found in many allelopathically active plants are dependent on the concentration of the active matter (Fig. 4). Linalool, cymol, citral, gaillardine and austricine excreted by many flowers, stimulated the pollen germination upto 10^{-7}-10^{-6} M concentrations and depress in higher ones. Rodriguez *et.al* (1976) described the biological effects of some sesquiterpene lactones, which are allelochemicals.

Table 3 Effects of volatile plant excretions and water extracts (1:1000 w/v) from allelopathic species with pesticide features on the pollen *Hippeastrum hybridum* germination.

Excretion	Quantity of pollens with pollen tubes formed (% of control)		
	Near slide	2.5 cm distant	Water extract from the same probe
Lavender (*Lavandula vera*) oil	0± 2	38 ± 6	–
Disrupted fruits (79 mg) of *Capsicum annuum*	55±10	108± 11	26.5 ± 8
Cutted bulb of garlic *Allium sativum* L. (350 mg)	98± 8	132± 10	73± 5

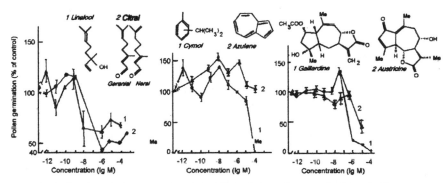

Fig. 4 Effects of volatile and liquid allelochemicals on the germination of *Hippeastrum hybridum* pollen

Experiment 3. *Effects of alkaloids – allelochemicals on the microspores germination.*

Alkaloids from allelopathically active species were also tested on the germination of vegetative and generative microspores. They are known as medicinal drugs and in most cases inhibition (30-80%) was observed (Fig. 5). Isoqinolinic alkaloids (berberine, glaucine and sanguinarine) inhibited the germination of vegetative microspores, but glaucine did not affect the pollen germination. Berberine inhibited the reactions only at low concentrations. The differences may be due to the different chemical structure - the number of five carbon-heterocycles and the position of benzene rings. Colchicine and casuarine have another chemical structures, but all these inhibited the germination of microspores.

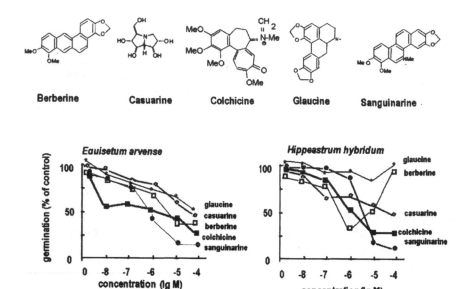

Berberine   Casuarine   Colchicine   Glaucine   Sanguinarine

Fig. 5 The effects of alkaloids -allelochemicals on the germination of vegetative microspores from *Equisetum arvense* and generative microspores (pollen) from *Hippeastrum hybridum*.

2.3   PISTIL STIGMA (FEMALE GAMETOPHYTE) OF SEED-BEARING PLANTS

Pistil is female reproductive organ of flower which is recipient of pollen. Flowers of knights'star (*Hippeastrum hybridum*) (Fig. 6) is suitable object to study the fertilization due to large sizes of pistil (10-15 sm) and pistil stigma (2-3 cm in diameter), variability in the time of flowering (round the year in green-house), high rates of fertilization (1-2 days after the pollination) and formation of fruits and viable seed during 1-1.5 months.

Principle: The observation of the rate and success of fertilization on the formation of fruits and seeds in the plant was used. The pistil stigma is treated with allelochemical and then pollen is added or pistil is not treated, but pollen treated with allelochemical is added on the pistil stigma (Roshchina, 2001 a, b).

Materials required: Flowering plants of knights'star (*Hippeastrum hybridum,*) vessels for allelochemicals for moistening the pistil or pollen with allelochemicals.

Procedure: The open pistil stigma (Fig. 7C) having a pleasant aroma (like linalool), was treated with allelochemical (the flow of needed solution on

Fig. 6 The common view (A,B) and fruit (C) of flowering knights' star (*Hippeastrum hybridum*)

pistil stigma or 2 min dip into the solution), and then pollen is added or pistil is not treated, but pollen preliminary treated with allelochemical is added on the pistil stigma. If the fertilization has occurred that is seen from the flower closing, the following observation of the fruit and seed formation takes place. The weight of fruits and seeds, amount of seeds and their viability (the formation of seedlings) are determined.

Observations: Observation on each plant is recorded for fertilization for 1-2 days after the pollination. Formation of fruits and seeds lasts during 1-1.5 months.

Statistical analysis: The data are analysed for the weight of fruits and seeds as well as amount of seeds formed.

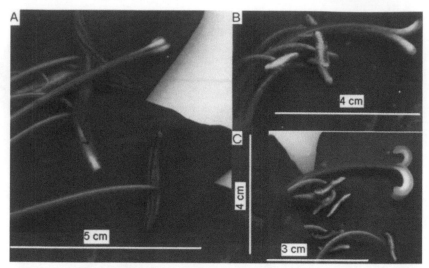

Fig. 7 The position of the pistil at the opening of the stigma

Experiment 1. *Effects of allelochemicals on the fertilization of fruits and seeds formation*

Fig. 8 shows how the 2 min treatment of pistil surface (see Procedure and Observation) with 10^{-5} M allelochemicals (before pollination occurred) influence the fruit formation and morphology. The open pistil stigma was treated with needed solution, which was put on pistil stigma or 2 min-dip into the same solution, and then pollen is added. Closing of flower in 1-2 days after the pollination shows that the fertilization has occurred (the pollen tube has achieved egg cell and fused with the cell). Fruits and seeds formed during 1-1.5 months. Alkaloids physostigmine, neostigmine (analogue of physostigmine), d-tubocurarine, yohimbine and sesquiterpene

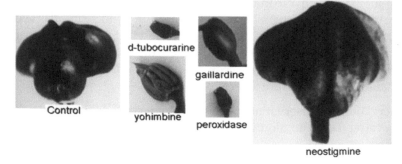

Fig. 8 The effects of allelochemicals on the fruit formation.

lactone gaillardine as well as peroxidase 3 mg/ml either blocked the fruit appearance or changed the weight and form of the fruits.

D-tubocurarine from *Chondrodendron tomentosum* Ruiz *et* Pav. and *Rauvolfia serpentina* Benth. *et* Kurz and horse radish peroxidase 3 mg/ml, which prevent the peroxide formation, block the fruit formation as a whole, while yohimbine and gaillardine inhibit the seed formation. Neostigmine stimulates the fruit and seed yield, although its precursor physostigmine has no significant effect (Roshchina and Melnikova, 1998).

3. ANALYSIS OF ALLELOCHEMICAL BINDING IN A CELL

3.1 Hydrophilic Compounds

If the allelochemical is hydrophylic, it cannot enter into the cell and act from outside by binding with chemoreceptors. The compounds from allelopathically active plants may serve as chemosignals and their signalling occurs via alternative pathways: (i) Chemoreceptor (sensors) $\rightarrow$ transducins (G-proteins) $\rightarrow$ secondary messengers (Ca^{2+}, cyclic AMP or GMP, inositol triphospate, etc) $\rightarrow$ organelles; or (ii) Chemoreceptor (sensors) $\rightarrow$ ion channels $\rightarrow$ action potential $\rightarrow$ organelles, or (iii) Chemoreceptor (sensors) $\rightarrow$ ion channels $\rightarrow$ cytoskeleton $\rightarrow$ organelles (Roshchina, 2005 a). What is the effect of acted allelochemical on the pathways, could be analysed to study the effects of substances on separate sites of the transduction chain.

Principle: Hydrophilic allelochemicals, mainly acetylcholine, biogenic amines, known as neurotransmitters and alkaloids, which are their agonists and antagonists, may bind with chemoreceptor(s) and sensors on the cell surface, then the chemosignal transfer within a cell (Roshchina, 2001 a). First response to the compounds was alterations in cellular fluorescence excited by UV (360-380 nm) light (see details in Chapter 9) as seen in Fig. 6, and activity of the surface sensor in particular cholinesterase histochemically stained with red analogue of Ellman reagent (Fig. 7) (Roshchina, 2001 a). More distant response to neurotransmitters and antineurotransmitter substances was the germination of plant microspores. To use appropriate inhibitors for the germination analysis, the target for allelochemical may be determined.

Materials required: Vegetative microspores of *Equisetum arvense* or pollen from *Hippeastrum hybridum* microspores, ordinary light microscope, luminescent microscope (may be combined with microspectrofluorimeter or laser scanning confocal microscope), Laser scanning confocal microscope, object glasses, Petri dishes, acetylcholine, biogenic amines dopamine, noradrenaline, serotonin and histamine, inhibitors (blockers) of ion channels -

a-bungarotoxin, blocker of sodium channels and antagonist of acetylcholine for nicotinic cholinoreceptor, tetraethylammonium, blocker of potasium channels and antagonist of acetylcholine for nicotinic cholinoreceptor, verapamil, blocker of calcium channels, fluorescent dye (Bodipy –dopamine), Red analogue of Ellman reagent, acetylthiocholine, dopamine. noradrenaline, serotonin,

Procedure: After the treatment of cell with allelochemical, the fluorescence of cellular surface is observed and then their germination is determined. The fluorescence and germination (see 3.1 and 3.2.) of the microspores was estimated by luminescent microscope or microspectrofluorimetry, or laser scanning confocal microscopy. The observation of a germinating microspore after 30 min of the treatment with tested allelochemicals was determined by luminescent microscope. If the allelochemical is able to fluoresce in blue or blue-green and act on the surface chemoreceptor, the blue or blue-green emission is seen only on the surface of the microspore (Fig.9). Another target of the allelochemical tested also may be a surface sensor - cholinesterase (Fig. 10), whose presence is determined for pollen *Hippeastrum hybridum* in ordinary light microscope as blue colour after 1 h exposure in 10^{-3} M of acetylthiocholine in 20 mM K- phosphate buffer pH 7.4 (Roshchina and Melnikova, 1998). If contractile proteins are affected with the allelochemicals, its fluorescence could be seen after the preliminary inhibition of the microspores with allelochemical (Fig.11) and the rate of the spores germination is also estimated.

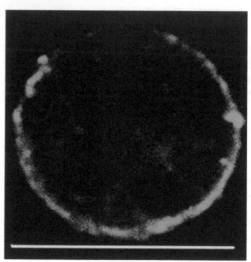

Fig. 9 The fluorescing surface of pollen from *Plantago maior* treated with 10^{-5} M of d-tubocurarine, which is concentrated in sites of cholinoreceptors. Bar = 20 mm.

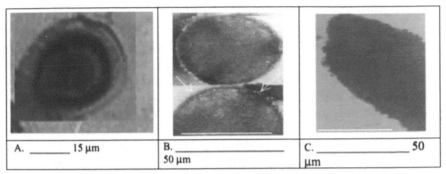

A. _____ 15 µm	B. _____ 50 µm	C. _____ 50 µm

Fig. 10 Histochemical staining of   *Hippeastrum hybridum* pollen (A), *Epiphyllium hybridum* pollen (B) and part of pistil of *Hippeastrum hybridum* (C) by red analogue of Ellman reagent. Blue color means the presence of cholinesterase.

Observations: After the cell treatment with allelochemical during 30 min-exposure, the fluorescence of cellular surface is observed. Mainly alkaloids binding with cholinoreceptors or adrenoreceptors demonstrate an increased emission in ultra-violet light of luminescent microscope as fluorescent probes for the study of cellular location of receptors. Besides fluorescence, 60 min exposure with acetylthiocholine 10^{-3}M and following 30 min-histochemical staining of cholinesterase as a surface sensor with Red analogue of Ellman reagent (Roshchina, 2001a) permits to see the blue color of the product of the substrate hydrolysis

Statistical analysis: In each treatment, 10 microspores were used to measure the maximal fluorescence. The means and their standard errors are determined, if the investigator has microspectrofluorimeter or microspectrofluorimeter having special statictic t- test programme.

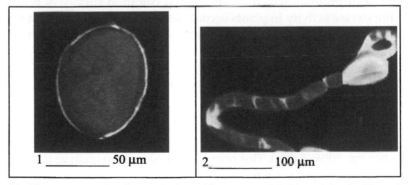

1 _____ 50 µm	2 _____ 100 µm

Fig. 11 Fluorescence of pollen grain from *Hippeastrum hybridum* stained with azulene 10^{-5} (left, bright lightening of cell wall and less intensive emitted nucleus in centre) or pollen tube stained with colchicine 10^{-7} M (right, lightening parts of pollen tube may be tubulin-binding sites).

Experiment 1. *Chemoreception on the cell surface*

After 30 min- exposure of microspores with the 10^{-6}- 10^{-5} M solution of compound studied (see Procedure and Observation) the fluorescence of the preparation was observed under luminescent microscope or analysed with both microspectrofluorimetry (see Chapter 9) and laser scanning confocal microscopy (see Chapter 8). The acetylcholine agonist (in particular muscarine from fungi *Amanita muscaria*) or antagonist d-tubocurarine (found in *Chondrodendron tomentosum* Ruiz et Pav. and *Rauvolfia serpentina* Benth. et Kurz), atropine (which is present in *Atropa belladonna* L.), yohimbine (found in *Corynanthe yohimbe* K.Schum), emitted mainly in blue (max. 450-470 nm) or in green (near 500-520 nm). Blue-green rings of the studied agonists or antagonists were seen in the luminescent microscope on the cellular surface (in particular for d- tubocurarine as shown in paper of Roshchina, (2005 b), where they were bound with proposed receptors. One of the examples are shown for the pollen of *Plantago major* treated with the fluorescing compound (Fig.9).

Experiment 2. *Cholinesterase as a sensor on the cell surface*

A target of the allelochemical may also be a surface sensor-cholinesterase (Fig. 10). If after the staining with Red analogue of Ellman reagent the blue colour is absent in the allelochemical treated microspore, possible target is the enzyme (Roshchina, 2001a,b) as for alkaloid berberine tested. If after the treatment by the test allelochemical, the colour is absent or light, the compound inhibits the enzyme (also see biochemical assay in Chapter 11).

Blue colour ring concentrated on plasmalemma of pollen (A). Arrows on B (lower part) shows a difference in the colour between cellular surface (blue color) and apperture for the output of pollen tube (red colour). Pistil excreted blue colour product, which covers the red coloured surface of pistil. The cholinesterase activity in plants is considered as sensitive test to study the allelopathic activity (Roshchina and Roshchina 1993; Roshchina,1999; 2001a).

3.2   Hydrophobic Compounds (Fluorescence of Allelochemicals in Cell)

Hydrophobic compounds such as sesquiterpene lactones and some alkaloids penetrate into cells and binds with various compartments (Roshchina, 2004; 2005b). This observation permits to analyze the possible target of the substance acted.

Principle: In the interaction with cell-acceptor fluorescent allelochemical may show the location of the target of the action.

Materials required: Vegetative microspores of *Equisetum arvense* or pollen from *Hippeastrum hybridum* microspores, ordinary light microscope, luminescent microscope

Procedure: If the fixed sample is studied, the cells are treated with tested allelochemical $10^{-5} - 10^{-4}$ M dissolved in organic solution (ethanol, chloroform, dimethylsulfoxide). But if the vital sample is needed, the solution is first diluted with water up to the ratio of organic solvent: water, was more than 100.

Observation: After 30-60 min of moistening in the solution, the cells are studied by luminescent microscopy.

Experiment 3. *Fluorescence and pollen germination at binding of allelochemicals with cytoskeleton* and organelles
Some allelochemicals such as sesquiterpene lactones or alkaloids penetrate into a cell, binding with various cellular compartments, and changing the cellular fluorescence excited by ultra-violet or violet light. This makes clear cellular mechanisms of actions for the allelochemicals. Sesquiterpene lactones azulene and proazulenes binds DNA-containing structures such as nuclei and chloroplasts, which fluoresce in blue (Roshchina, 2004).

Azulene can induce fluorescence not only DNA-binding structures in blue region of the spectra , but also the cell wall, which start to emit in orange-red (Fig. 11). The fluorescence of alkaloid colchicine from *Cochicium* genus or cytochalasin B from fungi *Helminthosporium dematioideum* is observed after the treatment of germinated pollen, whose production is depressed (Roshchina, 2005 a,b). Cochicine binds monomers of contractile protein tubulin that prevents the microtubules self-aggregation. The fluorescing parts of pollen grain and developing pollen tube are related to tubulin, which is marked with colchicine (Fig.11)

4. SUGGESTED READINGS

Golovkin, B.N., Rudenskaya, R.N., Trofimova I.A. and Shretter, A.I. (2001). *Biologically Active Substances of Plant Origin.* Nauka, Moscow, 1280 pp (3-volumes).

Muravnik L.E. and Ivanova A.N. (2002) Ultrastructural characteristics of leaf secretory cells from *Droseraceae* in relation to naphthoquinone synthesis. *Botanical Journal* **87** (11): 16-25. (In Russian).

Murphy, S.D. (1999). Is there a role for pollen allelopathy in biological control of weeds. In: *Allelopathy Update,* (Ed., S.S. Narwal) Vol. **2**: 321-332. Science Publishers, Enfield, Plymouth, USA.

Rodriguez, E., Towers, G.H.N. and Mitchell, J.C. (1976). Biological activities of sesquiterpene lactones. *Phytochemistry* **15**: 1573-1580.

Roshchina, V.V. (1999). Mechanisms of cell-cell communication. In: *Allelopathy* : *Basic and Applied Aspects*. (Ed. S.S. Narwal). Vol. **2**: 3-25. Science Publishers, Enfield, Plymouth, USA.

Roshchina, V.V. (2001a). *Neurotransmitters in Plant Life*. Science Publishers, Enfield, Plymouth, NH, USA.

Roshchina V.V. (2001b). Molecular-cellular mechanisms in pollen alllelopathy. *Allelopathy Journal* **8**: 11-28.

Roshchina V.V. (2002). Rutacridone as a fluorescent dye for the study of pollen. *Journal of Fluorescence* **12**: 241-243.

Roshchina V.V. (2003). Autofluorescence of plant secreting cells as a biosensor and bioindicator reaction. *Journal of Fluorescence* **13**: 403-420.

Roshchina,V.V. (2004). Cellular models to study the allelopathic mechanisms. *Allelopathy Journal* **13**: 3-16.

Roshchina V.V. (2005a). Contractile proteins in chemical signal transduction in plant microspores. *Biological Bulletin, Ser. Biol.* **3** : 281-286.

Roshchina V.V. (2005b). Allelochemicals as fluorescent markers, dyes and probes. *Allelopathy Journal* **1**: 31-46.

Roshchina VV (2005 c). Biosensors for the study of allelopathic mechanisms and testing of natural pesticides. In: Proceedings of International Workshop on Protocols and Methodologies in Allelopathy, April 2-4, 2004 Palampur. (Eds. G.L. Bansal and S.P.Sharma). Pp. 75-87. College of Basic Sciences, Palampur, India.

Roshchina, V.V. and Melnikova, E.V. (1998). Allelopathy and plant generative cells. Participation of acetylcholine and histamine in a signalling at the interactions of pollen and pistil. *Allelopathy Journal* **5**: 171-182.

Roshchina, V.V. and Mel'nikova, E.V. (1999). Microspectrofluorimetry of intact secreting cells applied to allelopathy. In: *Principles and Practices in Plant Ecology* : *Allelochemicals Interactions*, (Eds., Inderjit, K.M.M.Dakshini, and C.L.Foy). Pp.99-126. CRC Press, Boca Raton.

Roshchina, V.V. and Roshchina, V.D. (1993). *The Excretory Functions of Higher Plants*. Springer-Verlag, Berlin. 314 pp.

Roshchina, V.V. and Roshchina, V.D. (2003). *Ozone and Plant Cell.* Kluwer Academic Publishers, Dordrecht pp. 240.

Roshchina, V.V., Melnikova, E.V., Yashin, V.A. and Karnaukhov, V.N. (2002). Autofluorescence of intact spores of horsetail (*Equisetum arvense* L.) during their development. *Biophysics (Russia)* **47**: 318-324.

Roshchina, V.V., Miller, A.V., Safronova, V.G. and Karnaukhov, V.N. (2003). Reactive oxygen species and luminescence of intact cells of microspores. *Biophysics (Russia)* **48**: 259-264.

Roshchina, V.V., Yashin, V.A. and Kononov A.V. (2004). Autofluorescence of plant microspores studied by confocal microscopy and microspectrofluorimetry. *Journal of Fluorescence* **14** : 745-750.

Microalgae for Determing the Effects of Allelochemicals

C. Ciniglia, G. Pinto, A. Pollio, G. Aliotta[1] and P. Fergola[2]

1. INTRODUCTION

Algae are thallophytes (plants without roots, stem and leaves) that have chlorophyll *a* as their primary photosynthetic pigment and lack a sterile covering around the reproductive cells. This definition includes many of the plant forms that are not necessarily closely related (e.g. Cyanobacteria, are closer in evolution to the bacteria than to algae, or, on the contrary, the green algae are closer in evolution to the higher plants). Algae vary from small, single-celled forms to complex multicellular forms. They exhibit a wide range of reproductive strategies from simple, asexual cell division to complex forms of sexual reproduction. They occur in most habitats, ranging from marine and freshwater to desert sands and from hot boiling springs to snow and ice. They are extremely important ecologically, since they function in most habitats as the primary producers in the food chain, producing organic material from sunlight, carbon dioxide and water. Besides forming the basic food source for these food chains, they also produce the oxygen necessary for the metabolism of the consumer organisms. The algae probably account for more than half the total primary production

Dipartimento delle Scienze Biologiche, Sezione di Biologia Vegetale, Via Foria 223, I-80139 Napoli, Italy.
[1] Dipartimento di Scienze della Vita, Seconda Università di Napoli, Via Vivaldi 43, I - 81100 Caserta, Italy.
[2] Dipartimento di Matematica "Renato Caccioppoli", Università di Napoli Federico II, Via Cinthya, Napoli, Italy
E-mail: ciniglia@unina.it

worldwide, virtually all aquatic organisms are dependent on this production. Photoautotrophic organisms often develop severe competition for resource, space, light or nutrients. The release of allelopathic compounds which can interfere with settlement and growth of competitors, affecting population dynamics (Pratt, 1966; Rojo et al, 2000), is an adaptive strategy adopted by primary producers and microorganisms (Gross, 2003). This hypothesis was reported for the first time by Schreiter (1928), who cited a case, where the phytoplankton population of a pond was lowest in years, when the large aquatic plants were most dense.

In the following years, many researchers studied the 'interference' of macrophytae on the phytoplankton and their results showed that even if other explanations are possible, the allelopathic limitation of algal growth cannot be excluded. Allelopathy in marine systems may occur between phytoplankton species or in benthic areas where macroalgae, corals and few species of angiosperms are present. Marine environments are predominantly represented by phaeophytes, rhodophytes and chlorophytes. They establish different kinds of interaction with other photoautotrophs and also sometimes with heterotrophs. Allelochemical interference of macroalgae with microalgae has been observed for a long time (McLachan and Craigie 1964). The liberation of allelochemicals which prevent the growth of microalgae on macrophytes surface occurs frequently, as well as antifouling activity of macroalgae against bacteria and fungi (Hellio et al., 2000). Allelopathy could be one of the responsible factors that promotes the dominance of marine and freshwater harmful algal bloom (HAB)-forming species over the phytoplankton species. Some of these HAB species produce secondary metabolites, which can be toxic to other microorganisms and may function in grazing deterrence or as pheromones or as allelochemicals (Turner and Tester, 1997; Wyatt and Jenkinson 1997; Wolfe, 2000; Rengefors and Legrand, 2001; Schmidt and Hansen, 2001). The allelopathic studies on algae can be carried out either in the field or in the laboratory. This chapter mainly deals with laboratory tests, but some field methods will be also described.

2. FIELD INCUBATION STUDIES

Incubation of cultures *in-situ* in the sea or lakes, is the most common method. The advantage of field incubation or, *in-situ*, study is that one does not use 'standard conditions'of growth but relies on available nutrient supplies, as well as the temperature and light conditions prevailing at the time. The disadvantage is the "infinite range of variables found in nature" (Fogg, 1975). To plan *in situ* experiments on allelopathy among aquatic primary producers, it is necessary to know preliminarily the chemical, physical and biological features of the water body to be examined. For example as regards to algae, it is important to know the natural succession of the different species along the period we consider.

Moreover, the following points are also important.

I. The source of allelochemicals: The allelochemicals found in the water are produced by phycoplanktonic or bentonic algae, by aquatic macrophytes cohabiting with the algae, or are they present in the water body accidentally? (They can be released by plants living on the border of the water body or by leakage water etc.)

II. The fate of allelochemicals released in the environment: The allelochemicals under natural conditions can modify their chemical structure by abiotic (light, temperature, humic acids, etc,) or biotic agents (biotrasformations) and can produce a variety of different compounds.

2.1   Field Incubation Procedure and Experiments

Standard species can be incubated in dialysis tubes with the following procedure:

Seal one end of a 15 cm length of dialysis tubing, fill with distilled water

Insert and secure a short length of 5 mm glass tubing, plugged with cotton, into other end

Autoclave, submerge in a beaker of water

Inoculate via glass tube with axenic cultured cells, remove tube and seal dialysis tubing

Place array of dialysis cultures as desired *in situ*, for days to weeks (depends on fouling rate)

This procedure can be also implemented by using more than one algal species, which coexist in the habitat under examination. These are the so-called **Community dynamics in in-situ enclosures**. The enclosures containing either single species or algal mixed populations can be placed at different depths and the effect of allelochemicals present in the water body are measured on the growth of each species, by monitoring through time (hours to years) under replicated experimental manipulations.

3. LABORATRORY STUDIES

3.1   Continuous Culture Systems for Macrophytes and Microalgae

The occurrence of toxic compounds in plant tissues is not necessarily related to allelopathy. Allelopathy should be evidenced through experiments in which a toxic product is shown to be released from the putative aggressor, and arrives at the putative victim in functional concentrations under reasonably natural conditions (Inderjit and Callaway 2003). First of all, laboratory experiments dealing with allelopathy should demonstrate the release of a compound in the medium. Two methods to collect allelchemicals released by laboratory cultures of macrophyte or microalgae are described in Sections 5 and 6.

3.2   Continuous Culture Methods for Microalgae

Continuous culture systems have been widely used to culture microorganisms for industrial and research purposes (Kubitschek 1970; Tempest 1970; Veldkamp 1976; Rhee 1980). In recent years, these culture techniques have found their way into the bioassay methods of ecotoxicology and allelopathy (Rhee 1980). The early development of a continuous culture system can be traced back to the work of Novik and Szilard (1950 a,b) who developed the first 'chemostat'. In a continuous culture system, nutrients are supplied to the cell culture at a constant rate and to maintain a constant volume, an equal volume of cell culture is removed. This allows the cell population to reach a 'steady state', where the growth rate and the total number of cells/ml of culture remains constant. Two kind of continuous culture systems can be distinguished: 'turbidostat' and 'chemostat'.

I. Turbidostat: In the turbidostat system, fresh medium is supplied only when the cell density of the culture reaches some predeterminated level, as measured by the extinction of light passing through the culture. At this point, fresh medium is added to the culture and an equal volume of culture is removed.

II. Chemostat: In a chemostat, the medium is delivered at a constant rate, to keep constant growth rate. The nutrient medium is supplied to the culture vessel at a constant rate by a peristaltic pump used to control the washout rate. The rate of media flow is often set at approximately 20% of culture volume per day. Air is pumped into the algal culture vessel through an air compressor controlled by a flow meter and carried in two flasks of sterile water. This bubbling air has three effects: (i) it supplies CO_2 and O_2 to the culture, (ii) allows circulation and agitation of the cultures and (iii) pressurizes the head space of the culture vessel to provide the force to 'remove' an amount of media (and cells) equal to the volume of inflowing media. The magnetic stirrer and aeration help to prevent the cells from collecting in the bottom of the culture vessel. The outflow can be collected and filtered and the filtrate is tested on other organisms to determine their hypothetical allelopathic effects (Fig. 1).

Allelopathic interactions between two algal species can be also studied by inoculating both algal species, previously grown in different flasks, in the same culture vessel in a 1:1 ratio (cross-culturing). The allelochemicals liberated in the culture medium can affect the growth of one another, but in this approach it is hard to discriminate among allelopathic effects and removal or reduction of some nutrient (exploitative competition,). General mathematical models for them have been proposed (see Section 3.7).

3.2.1 Continuous trapping apparatus

Some potential allelochemicals for studies against the microalgae could be isolated from aquatic plants and from macroalgae. In this approach, macrophytes can be cultured in a modified continuous trapping apparatus, containing sterile culture medium. The apparatus is kept at constant temperature (26-28°C) and illuminated with fluorescent lamps under a 16-8h light-dark photoperiod. The nutrient medium is supplied to the apparatus at a constant rate by a peristaltic pump and collected through an Amberlite column situated at the base of the apparatus. This column can be sequentially eluted with polar sovent (methanol, acetone, chloroform) and all the fractions isolated and identified by chemical analytic techniques and successively used for biological assays.

3.3 Algal Assays to Monitor Alellochemicals Toxicity

Laboratory algal assays for monitoring the toxicity of an allelochemical are very useful, due to the ease and rapidity with which the organism can be cultured and their responses measured. They also require less equipment, space and time of the researcher.

3.3.1 Two-step approach

The two-step approach is based on the combined use of two different techniques.

First of all, the allelopathic activity of a compound (or of a mixture of compounds) isolated as per the above methods should be tested in agar diffusion assays using different algal strains The selection of an assay alga depends primarily on its ability to grow satisfactory under the chosen experimental conditions. Algae can be selected either on the basis of their taxonomic position (i.e. algae belonging to different algal classes or orders) or for their diffusion over a wide array of aquatic habitats.

I. Filter paper disk bioassays

Selection of different algal strains and their cultivation in liquid medium
↓
Algal cultures in exponential phase are used for inoculating
Petri dishes containing agarized culture medium
↓
Paper disks (6 mm diameter) are impregnated with organic solvents or aqueous solutions containing a single allelochemical or a mixture
↓
If organic solvents are used extract impregnated discs are dried in a current of warm air to allow solvent evaporation,
↓
Paper disks are placed on the inoculated agar surface
↓
Inoculated Petri dishes are kept under fixed conditions of light and temperature for seven days
↓
The inhibitory action of the allelochemical is measured at the end of the experiments as the diameter of no growth zone around the paper disk.

The main advantage of this technique is the rapidity of response and the possibility of performing many bioassays with different algal strains at the same time. The disadvantage is that the kind of response is qualitative. This limitation can be overcome by selecting for liquid inhibition bioassays, the strains that give a positive response in paper disk bioassay (i.e. the strains sensitive to the allechemical(s) tested).

II. Bioassay in a liquid medium: In the bioassays on liquid medium the test alga is inoculated into a liquid medium containing the allelochemical. Some of the following important aspects should be considered in the experiment design:

(i) The appropriate size of flasks is proportional to the duration of the experiment and to the total volume of samples to be removed each day for measurements of growth.
(ii) Levels of nutrients in the culture medium should be 'realistic', no more than 2-3′ probable natural levels, to avoid toxic effects or inducing limitation by another nutrient.

(iii) The substances to be tested can be added in two ways:

A. Spiked: Is the compound or the mixture that is added at the beginning of experiments.

B. Fixed level or supply rate: Consists of periodic or continuous additions to prevent depletion of the compound. A fixed level can be maintained only when the concentration of the allelochemical can be measured during the course of the experiments. On the other hand, a fixed supply rate does not require concentration measurements.

With liquid media bioassays, it is possible to obtain a lot of information about the activity of the allelochemical (toxicity, type of inhibition on the growth curve, morphological modification on the test alga, etc.). But the time of response can be long (about 15 d) and for disk bioassay more equipment, space etc. are necessary.

III. End Point test: The method for testing the toxicity of a compound (end point test) has been developed for the green microalga *Pseudokirchneriella subcapitata* (EPA, OECD), but can be extended to a wide range of microalgae. The method is summarized below:

Grow algal stock culture in standard medium, at standard light and temperature conditions.

↓

Axenically dissolve the allelochemical in sterilized culture medium

↓

Inoculate 49 ml medium with the allelochemical(s) in 100 ml flask and add 1 ml from the stock culture of each strain to have in the flask a final concentration of 10^4 cells×ml^{-1}

↓

Prepare also a blank containing no allelochemical in the culture medium

↓

Different concentrations of the allelochemical should be tested and for each concentration (and for the blank) at least three replicates should be prepared

↓

Grow for 72-96 h, then count cells

4. MEASUREMENT OF RESPONSE

(i). The purpose of the experiment dictates the appropriate response variables. Ideally, at least two independent measures of response should be used. Generally often practical constraints such as experimental container size dictates the sampling regime (size, frequency, number of replicates and the kinds of measurements possible). It is critical to have **initial estimates** of each response variable from each experimental unit, especially when species composition is involved. This is because phytoplankton may clump and not be evenly distributed among treatments, both quantitaively and qualitatively.

(ii). Biomass/standing crop is most readily estimated from Chl a, usually by fluorometry.

(iii). Biomass also can be estimated as particulate C, N or P or (rarely) protein, carbohydrate or fat.

(iv). More tedious but most informative is cell counting (either by automated particle counter or preferably by microscopic observation).

(v). In some cases, spp. composition may change, this may not be unambiguously evident in Chl or particle count data.

(vi). For a more carbon-energy-flow type parameter, cyctoplasmic (not cell) volume and cell carbon may be estimated.

(vii). Rate measurements such as photosynthesis (^{14}C uptake) or nutrient uptake are also often used, although these are susceptible to variable short term physiological state (diel rhythms, cell cycle stage, etc.) that may be unrelated to the water quality parameters of interest, and therefore may be misleading. Also, they are essentially an instantaneous response, which may be qualitatively and quantitatively different from an integrated long term response. The delay time between treatment initiation and measurement of response is an important consideration.

(viii). Population growth rate is a better (longer term) parameter, but it is rarely feasible for individual species in mixed assemblages and the community average rate obscures spp. shifts.

(ix). Note that in some cases, it should not be inferred that no response to a particular nutrient means that it is irrelevant to management. For example, excessive anthropogenic P loading may ↑ algal biomass tremendously until some other element becomes 2° limiting; in this case substantial ↓ P loading would ↓ algal biomass even though assays indicate excess P.

4.1 Response Patterns

Generally, it is possible to identify many response patterns of alga to allelochemicals, but the following are the most common:

(i). A gradual increase in sensitivity to the allelochemical with a larger concentration.

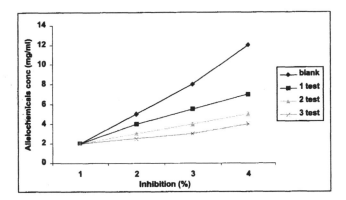

(ii). initial sensitivity to the allelochemical with eventual resistance and growth after the induction period.

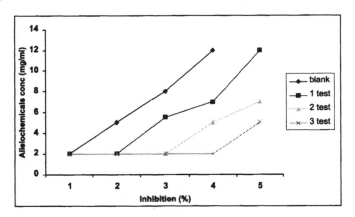

(iii). Resistance even at high concentration, followed by sensitivity with a subsequent dying off of the algal cells.

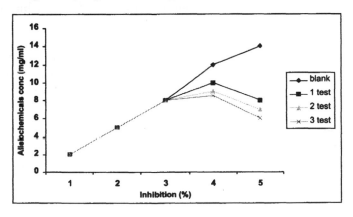

(iv). Complete resistance to the allelochemical

(v). Mild stimulation of algal growth over a limited range of low concentrations of the allelochemical

(vi). More marked stimulation of growth at a higher concentration of the allelochemical. We can also have other effects viz., (i) bio accumulation, (ii) biotransformation of the allelochemical.

4.2   Effects of Allelochemicals on Algae

The allelochemicals greatly influence the growth and development of aquatic plants including algae, through various biochemical and physiological interactions. The microscopic observations show the following effects of allelochemicals on algae :

a). Cell rupture

b). Variation in the cell size

c). Increase or reduction in spore number in the sporangia

d). Morphological alteration of organelles structure (mainly of chloroplasts)

e). Occurrence of cytoplasmatic and/or vacuolar deposits

4.2.1   Regreening of red colour cells of Chlorella zofingensis

Chlorella zofingensis cells are transferred to a growth medium, with a low nitrogen concentration (10% of normal concentration). After approximately 6-8 wk they develop a red colour, due to the decomposition of chlorophylls and synthesis of secondary carotenoids (stored in lipid droplets within the cytoplasm of the cells). At this stage the chloroplast are intact, although the surface area of thylacoids is mostly reduced.

If the culture is kept at 8°C, without aeration, the algae are physiologically stable for at least 6 mont. The course and quality of regreening process are not altered and remain reproducible. The algae used for the bioassay are harvested by centrifugation and suspended on a N-sufficient medium. The increase of chlorophyll synthesis leads to a complete regreening of the culture within seven d. During this period the total cell number remained approximately constant. The addition, toxicant in culture medium interferes with the regreening process, causing an inhibition of this process which is proportional to the toxicity of the compound.

Advantages

(i). The test can be used universally.

(ii). Standardized algal material is always available to conduct investigations.

(iii). The phytotoxicity of a substance can be determined at the same cell density for a long experimental period.

(iv). A reduction in the allelochemical toxicity against the algae due to the growth of cultured organism, which can decrease the toxicity, does not occur in this bioassay.

(v). The sensivity of this test is significantly higher than other toxicity-measuring systems.

4.2.2 SCGE technique to assess genetic toxicity of allelochemicals

In the last decade, the comet assay or SCGE developed by Singh (1988), is a very sensitive method for detection and quantification of DNA damage with applications in genotoxicity testing, human biomonitoring and molecular epidemiology and in ecogenotoxicology (Collins, 2004). The COMET assay attracts by its simplicity, versatility, speed and economy and the number of publications on SCGE grow year per year. The advantages of this technique are: (i) the collection of data at the level of the indivisual cell, allowing for better statistical analyses; (ii) the need for small numbers of cells per sample (< 10.000); (iii) its sensitivity for detecting DNA damage and (iv) any eukaryotic cell population can be analyzed.

The potential genotoxicity of allelochemicals has never been investigated. In this regard, we suggest the adoption of SCGE technique, a working protocol to detect the DNA damage is given below:

Grow algal stock culture in standard medium, at standard light and temperature conditions.
↓
Axenically dissolve the allelochemical in sterilized culture medium
↓
Inoculate 49 ml medium with the allelochemical(s) in 100 ml flask and add 1 ml from the stock culture of each strain to have in the flask a final concentration of 10^4 cells×ml^{-1}
↓
Prepare also a blank containing no allelochemical in the culture medium
↓
Different concentrations of the allelochemical should be tested and for each concentration (and for the blank) at least three replicates should be prepared
↓
Grow for 2h-96 h, then perform COMET assay
↓

Treated and untreated cells are collected by centrifugation and embedded in a three layered microgel, on a fully frosted microscopic slide, composed as listed below (i). a bottom layer of 1% normal melting agarose; (ii). a second layer of 0.5% low-melting agarose containig 50 ml of tested and untested algal solutions (iii). an upper layer of 0.7% low-melting agarose.
↓

The slides are dippped into a lysis solution containing 300 mM NaOH, 30 mM Na2EDTA and 0.01% SDS for 1 h
↓

The slides are dipped in an electrophoresis buffer (300 mM NaOH, 30 mM Na2EDTA) for 15 min at 4°C, to allow unwinding DNA.
↓

Electrophoresis are performed using the same buffer at 4°C for 20 min at 25V and 300 mA.
↓

Gels are neutralized embedding the slides twice in 400mM Trs buffer (pH 7.5)
↓

DNA molecules on gel are stained with ethidium bromide and observed using epi-fluorescent microscopy.
↓

DNA damage are analyzed with an automatic analyzing system which gives a measurement of DNA tail length and of head DNA %.

5. MATHEMATICAL MODELS

Mathematical models can be usefully applied to better understand the complex phenomenon of the allelopathic competitions, frequent in nature not only between algal species, but also between algae and bacteria, bacteria and bacteria, as well as algae and aquatic plants. To simplify both the model and the mathematics necessary for its investigation, we will study this phenomenon in a laboratory context, where many simplifications can be easily obtained. Suitable laboratory's devices, in fact maintain constant temperature, light, pH, and keep the nutrients solution homogeneous by stirring it. Under these conditions, the competition can be well represented by a 4 ODE's system, where the fundamental unknowns are the concentrations of nutrient (denoted by S), of populations (denoted by $N1$ and $N2$) and of allelochemicals (denoted by p).

A crucial role in the modelling and in the dynamics exhibited from the corresponding systems is played by the choice of functions which describe the following four main features:

(i). The nutrients uptake rates

(ii). The yield (ratio between the concentrations of produced biomass and absorbed nutrient)

(iii). The allelochemicals production;

(iv). The inhibitory or lethal effect of allelochemicals

When it is possible this choice should be performed on the basis of data we are able to extract from suitable laboratory experiments. In the other cases it will be suggested from our knowledge of the phenomenology .

The main choice of these functions should be based on data from suitable laboratory experiments. This approach has been recently applied to the mathematical study of the allelopathic competition between *Pseudokirchneriella subcapitata and Chlorella vulgaris* in an open culture (a chemostat-like device), where the nutrient is a mixture of inorganic phosphates Pi. Here we will illustrate step by step the four mentioned features, observing that each equation of the mathematical model can be actually constructed as a balance equation.

(i). Experiments done both in batch and open culture show that the two algal species exhibit different nutrient uptake rates and precisely *Pseudokirkneriella subcapitata* follows the non-monotone Adrews model (depending on three parameters), whereas, *Chlorella vulgaris* follows the monotone Michelis–Menten model (depending on two parameters).

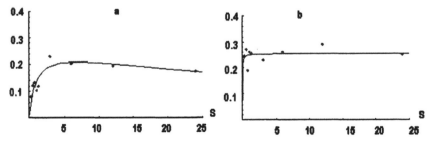

Fig. 1 a and b show how the Andrews and the Michaelis-Menten functions (with the optimal parameter values) respectively fit the experimental data

(ii). Based on the results of several experiments done both in batch and chemostat cultures, the yields of both species are constant and that *C. vulgaris*'s yield is almost twice than *P. subcapitata*.

(iii). The production of the toxic chemical compounds depends on the concentration of the producer species and that it has a cost. Precisely, this cost is paid by reduction in the growing potential. In other words, the overall energy coming from the use of nutrients (usually completely available for the growing process), in this case is partially diverted to the toxicants production. A linear function of the concentration of the producer species can be then introduced, to represent the regulating mechanism of such a production. This introduces a new parameter (denoted by k).

(iv). The experiments, show that the effect of Chlorellin on *P. subcapitata* is inhibitory. In particular, such inhibition decreases the exponential function, which introduces a new parameter (denoted by g).

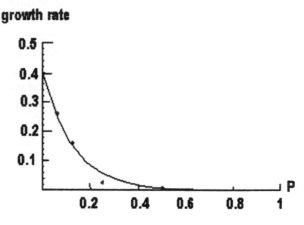

growth rate

Fig. 2 Dependence of *P. subcapitata* growth rate on the Chlorellin concentration p. Comparison between experimental data and the exponential function $f_1(S)\ e^{-\gamma p}$, where $\gamma = 7.81447$.

These considerations suggest the following mathematical model

$$
\begin{cases}
\dfrac{dS}{dt} = (S^0 - S)\,D - f_1(S)\,e^{-\gamma p}\,\dfrac{N_1}{\eta_1} - f_2(S)\,\dfrac{N_2}{\eta_2} \\[2mm]
\dfrac{dN_1}{dt} = N_1\,[f_1(S)e^{-\gamma p} - D] \\[2mm]
\dfrac{dN_2}{dt} = N_2\,[(1 - k\,N_2)f_2(S) - D] \\[2mm]
\dfrac{dp}{dt} = k\,N_2^2\,f_2(S) - Dp
\end{cases}
$$

where,

$S(t)$ the Pi concentration at time t;

$N_1(t)$ the concentration of *P. subcapitata* at time t;

$N_2(t)$ the concentration of *C. vulgaris* at time t;

$p(t)$ the concentration of toxicant at time t;

$\eta_1; \eta_2$ the constant yields of populations N_1 and N_2;

S^0 the constant concentration input of the limiting nutrient ($S^0 > 0$);

D the constant washout rate ($D > 0$);

m_1 the maximal specific growth rate of *P. subcapitata* ($m_1 > 0$);

m_2 the maximal specific growth rate of *C. vulgaris* ($m_2 > 0$);

γ a measure of the inhibiting effect of allelochemicals (chlorellin) ($\gamma > 0$);

k indicates the fraction of potential growth devoted to the production of allelochemicals, $0 < kN_2 < 1$.

$S(t)$ the Pi concentration at time t;

$N_1(t)$ the concentration of *P. subcapitata* at time t;

$N_2(t)$ the concentration of *C. vulgaris* at time t;

$p(t)$ the concentration of toxicant at time t;

$\eta_1; \eta_2$ the constant yields of populations N_1 and N_2;

S^0 the constant concentration input of the limiting nutrient ($S^0 > 0$);

D the constant washout rate ($D > 0$);

m_1 the maximal specific growth rate of *P. subcapitata* ($m_1 > 0$);

m_2 the maximal specific growth rate of *C. vulgaris* ($m_2 > 0$);

γ a measure of the inhibiting effect of allelochemicals (chlorellin) ($\gamma > 0$);

k indicates the fraction of potential growth devoted to the production of allelochemicals, $0 < kN_2 < 1$.

Moreover

$$f_1(S) = \frac{m_1 S}{a_1 + S + h_1 S^2} \text{ and } f_2(S) = \frac{m_2 S}{a_2 + S}$$

are the functional responses (Andrews for *P. subcapitata*, Michaelis-Menten for *C. vulgaris*) of the two populations.

System (1) is of the same type as other systems which have recently appeared in the literature on allelopathic competitions (Hsu and Waltman, 1998, Braselton and Waltman, 2001, Fergola et al., 2004). Usually the mathematical analysis of such systems, first requires one to check if their solutions satisfy the properties of global existence in the future, positivity, boundedness and uniqueness. Subsequently, it often happens that, due to the difficulties of integration of these systems, we look for special biologically meaningful solutions such as steady-states or periodic solutions. The methods of the qualitative analysis are then applied to get information on their stability properties. In particular, an extensive analysis of these properties is often carried out, by using both the linearization principle and the Liapunov Direct Method. Finally, several numerical simulations are performed by using the parameters of optimal numerical values, computed by applying some statistical approach to the experimental data (e.g. the non linear regression method). These simulations allow us to illustrate both the analytical results and to test new possible scenarios for forecasting further behaviours.

6. REFERENCES

Braselton, J.P. and Waltman, P. (2001). A Competition model. *Mathematical Biosciences* **173**: 55-84.

Collins, A. R. (2004). The COMET assay for DNA damage and repair. *Molecular Biotechnology* **26**: 249-261.

Fergola, P., Aurelio, F., Cerasuolo, M. and Noviello, A. (2004). Influence of mathematical modelling of nutrient uptake and quorum sensing on the allelopathic competitions. Proocedings "WASCOM 2003". *12th Conference on Waves and Stability in Continuous Media. World Scientific Publishing, River Edge, New Jersey*, pp. 191-203.

Fogg, G.E. (1975). *Algal Cultures and Phytoplankton Ecology*. 2nd. edition. University of Wisconsin Press, Madison, USA.

Gross, E.M. (2003). Allelopathy of aquatic autotrophs. *Critical Reviews in Plant Science* **22**: 313-339.

Hellio, G.B., Bremer A.M., Pons, G., Cottenceau, Y. and Le Gal, N. (2000). Antibacterial and antifungal activities of extracts of marine algae from Brittany France. Use as antifouling agents. *Applied Microbiology Biotechnology* **54**: 543-549.

Hsu, S.B. and Waltman, P. (1998). Competition in the chemostat when one competitor produces a toxin. *Japan Journal of Industrial and Applied Mathematics* **15** : 471-490.

Inderjit and Callaway, R.M. (2003). Experimental designs for the study of allelopathy. *Plant and Soil* **256**:1-11.

Irmer, U., Heuer, K. and Weber, A. (1985). Effects of various organic chemicals on the regreening of red colored *Chlorella zofingiensis*. *Ecotox and Environmental Safety* **9**:121-133.

Kubitschekk, H. E. (1970). *Introduction to Research With Continuous Cultures*. Prentice-Hall, Englewood Cliffs, NJ,USA

McLachlan, J. and Craigie, J.S. (1964). Algal inhibition by yellow ultraviolet absorbing substances from *Fucus vesiculosus*. *Canadian Journal of Botany* **42**: 287–292.

Novicik, A. and Szilard, L. (1950). Experiments with the chemostat on spontaneous mutations of bacteria. *Proceedings National Academy of Sciences* USA **36**: 708-719.

Novicik, A. and Szilard, L. (1951). Genetic mechanisms in bacteria and bacterial viruses. I. Experiments on spontaneous and chemically induced mutations of bacteria growing in the chemostat. *Cold Spring Harbor Symposia on Quantitative Biology* 16:337-343.

Pratt, D.M. (1966). Competition between *Skeletonema costatum* and *Olithodiscus luteus* in Narragansett Bay and in culture. *Limnology and Oceanography* 11: 447–455.

Rengefors, K. and Legrand, C. (2001). Toxicity in *Peridinium aciculiferum* - an adaptive strategy to outcompete other winter phytoplankton. *Limnology and Oceanography* 46: 1990-1997.

Rengefors, K., Ruttenberg, K.C., Haupert, L.C., Taylor, C., Howes, B.L. and Anderson, D.M. (2003). Experimental investigation of taxon-specific response of alkaline phosphatase activity in natural freshwater phytoplankton. *Limnolology and Oceanography* 48: 1167–1175.

Rhee, G.Y. (1980). Continuous culture in phytoplankton ecology. In: *Advances in Aquatic Microbiology*, M.R. Droop and H.W. Jannarch (eds). Vol 2 : 151-203, Academic Press, New York, USA.

Rojo, C., Ortega-Mayagoitia, E., Rodrigo, M.A. and Alvarez-Cobelas, M. (2000). Phytoplankton structure and dynamics in a semi-arid wetland National Park Las Tablas de Daimiel (Spain). *Archives of Hydrobiology* 748: 397–419.

Schreiter, T. (1928). The effect of *Elodea* on the growth of microplankton in Hirschberger pond in Bohemia during the years 1921-1925. Sborn´ýk vyzkumnych´ustavu zemedelskych rcs. V. Praze.

Schmidt, L.E. and Hansen, P.J. (2001). Allelopathy in the prymnesiophyte *Chrysochromulina polylepis*: Effects of cell concentration, growth phase and pH. *Marine Ecology Progress Series* 216: 67-81.

Singh, N.P., McCoy, M.T., Tice, R.R. and Schneider, E.L. (1988). A simple technique for quantitation of low levels of DNA damage in individual cells. *Experimental Cell Research* 175: 184-191.

Tempest, D.W. (1970). Theory of the chemostat. In: *Methods in Microbiology* ., J.R. Norris and D.W.Ribbons(eds). p 259-276. Academic Press, New York,USA.

Turner, J.T. and Tester, P.A. (1997). Toxic marine phytoplankton, zooplankton grazers, and pelagic food webs. *Limnology and Oceanography* 42:1203–1214.

Veldkamp, H. (1976). *Continuous Culture in Microbial Physiology and Ecology.* Meadowfield Press, Durham, UK. 25. Wolfe, G.V. (2000). The chemical defence ecology of marine unicellular plankton: constraints, mechanisms and impacts. *Biology Bulletin* 198: 225–244.

Wyatt, T. and Jenkinson, I.R. (1997). Notes on *Alexandrium* population dynamics. *Journal of Plankton Research* 19: 551–575.

Chapter

Bryophytes to Test
Allelochemicals *in vitro*

Simonetta Giordano

1. INTRODUCTION

The bryophytes are involved in many kinds of biological competition (Rice, 1979; 1984), such as bryophyte/bryophyte competition even between the plants of same species (self-incompatibility), bryophytes/ vascular plants, bryophytes/algae and especially due to their ecological similarity, bryophytes/lichens (During and Van Tooren, 1990). Lichens produce numerous secondary metabolic compounds, among which lichen acids are produced in a large quantity. These can be found in high concentrations (25%) but generally they are 5-10% of thallus dry weight (Lawrey,1995; Gardner and Mueller, 1981). They accumulate as crystals on the external surface of medullary hyphae, hence, available immediately as toxic compounds in biological competition.

Bryophytes are simple land plants which occupy various habitats and have a peculiar life cycle with haploid generation (gametophyte) prevailing over diploid phase (sporophyte). During the life cycle, mature gametophore is preceded by a filamentous stage of development, the protonema. This is a developmental stage of moss life cycle, that occurs both after meiospore germination and during vegetative reproduction (gemma germination) or regeneration by excised parts of gametophyte or even sporophyte (Fig. 1).

Dipartimento di Biologia Strutturale e Funzionale, Complesso Univesitario Monte S. Angelo, Via Cinthia, 80126 Napoli, Italy, E-mail: giordano@unina.it

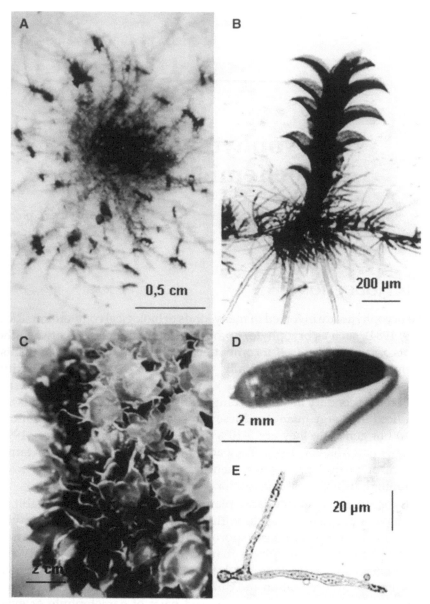

Fig. 1 Some phases illustrating moss life cycle. A - Regeneration *in vitro* from shoot tissue in *Tortella flavovirens*. The shoot in the middle (arrow) has produced a protonemal system from which new shoots have been formed in 2 months. B – Newly regenerated shoot. C – Shoots (gametophyte, haploid generation) of *Bryum capillare*. D – Capsule (sporophyte, diploid generation) of *Bryum capillare* that produces aploid spores by meiosis. E – A germinating meiospore producing primary protonema from which new shoots can arise. In A and C bar is 0.5 cm.

Moss protonema consists of three different cell types (Figure 2):

(i). **Chloronema:** The chloronema formed by short cylindrical cells with transverse septa, rich in round-ovoid chloroplasts representing the photosynthetic organ.

(ii). **Caulonema:** The caulonema, pale green, formed by longer cells with oblique septa, a few spindle-shaped plastids and with age, brown pigmented walls, involved in photosynthesis, metabolite transport and formation of shoot gemmae.

(iii). **Rhizonema:** The rhizonema is very similar to caulonema, the main difference being in the regular branching pattern with successive orders having narrower diameters, involved in anchorage, absorption, solute transport, tuber formation, with positive geotropism and thigmotropism (Schumaker and Dietrich 1997, Duckett et al. 1998), capable of being modulated by exogenous or endogenous stimuli.

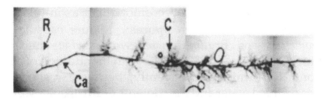

Fig. 2　Protonema of *Tortella flavovirens* obtained from shoot, after 2 months in culture. Light microscope x 50.　Ca= caulonema; C= chloronema; R=rhyzoids

In thalloid liverworts, spore germination produces a protonemal tube that stops growing very soon; from this short tube a plate is formed that develops into mature thallus. Many species can also vegetatively reproduce by propagules, these are multicellular diaspores directly developing into the mature thallus. Unlike the moss protonema, rhizonema-like filaments of liverworts are, in general, single celled.

There are many functions of protonema: it represents an organ of biosynthesis, substrate exploitation, absorption and solute transport and lastly of new shoot production. Whether it is produced during a sexual or vegetative cycle, it is a cellular clone that vegetatively multiply the number of new gametophores .

2. TEST OF CO-EXISTENCE IN VITRO

Experiment:　Field observations on cryptogamic vegetation dynamics in a Mediterranean, coastal macchia site, induced us to investigate possible interactions between the lichen and moss population mediated by allelochemicals (Giordano et al., 1999). In this experiment, the toxic effect induced by the lichen *C. foliacea* on the growth of some moss species was evaluated *in vitro* by a co-existence test.

2.1 Materials Required

Basic equipment for plant culture media preparation and sterilization, controlled growth chamber, microscope, stereomicroscope and inverted microscope equipped with a photocamera, laminar flow cabinet.

(i). **Mohr medium:** pH 7.5 (KNO_3 100 mg, $CaCl_2 . 4H_2O$ 10 mg, $MgSO_4$ 10mg, KH_2PO_4 136 mg, $FeSO_4$ 0.4 mg and 1 ml of BBM (Bold Basal Medium) solution (Nichols, 1973) volume made to 1000 ml with distilled water (Krupa, 1964).

(ii). **Plant material:** Field-grown lichen (i.e. *Cladonia foliacea*) and mosses (i.e. *Funaria hygrometrica, Bryum dunense, Barbula convoluta, Tortella flavovirens,* and *Pleurochaete squarrosa)* and liverwort (i.e. *Lunularia cruciata,* a widely-distributed, cosmopolitan, toxin-tolerant species).

(iii). **Protonemal system from spores:** Mature capsules of *Funaria hygrometrica* and *Bryum dunense* were surface sterilized in 70% ethanol (2 min) and in 2% NaClO with the addition of a few drops of Triton X-100 (5 min). Subsequently, they were washed (10 min) with sterile distilled water and the contents of 10 capsules were suspended in 10 ml of sterile distilled water. Samples (200 µ l) of this spore suspension were inoculated in modified Mohr medium. The spore cultures were kept in a climatic room at 18 ± 1°C, 70% constant relative humidity and 16 h light (2000 to 5000 lux)/8 h dark photoperiod and observed after 3, 7, 14 and 21 days. Spore germination (%) and protonemal growth were evaluated in aseptic cultures added or not (control) with lichen *Cladonia foliacea* extract (see later).

(iv). **Protonemal system from shoots:** In *Funaria hygrometrica* and *Bryum dunense,* 14 days after culture some control protonemata from spores were transplanted to solid (agar 1.5%) Mohr medium to obtain (in 2-3 months) stock culture of shoots. Thereafter, shoots were used to test the potential allelochemicals by cutting 0.5 cm pieces containing at least one shoot node. The shoots were sub-cultured in Petri dishes in liquid or solid medium to obtain secondary protonema.

(v). **Protonema regeneration from wild moss and liverwort:** Shoots of *Barbula convoluta, Tortella flavovirens* and *Pleurochaete squarrosa,* spp. (that only sporadically produce sporophytes), were collected and cleaned from dust and dead parts, thoroughly washed in tap water and then in solution of Triton X-100 (0.8%), rinsed in distilled water and surface sterilized for 10-20 seconds in 1% sodium hypochlorite solution and rinsed four times in sterilized distilled water.

Cultures were obtained by cutting the stems of *Barbula convoluta* and *Tortella flavovirens* into 0.5 cm pieces containing at least one shoot node. In *P. squarrosa* detached leaves were used avoiding very old and very young ones (Giordano et al., 1996). For *Lunularia cruciata*, round multicellular propagules contained in hemispheric cup carried on dorsal surface were used without any further treatment.

Plant material was transferred into Petri dishes in liquid Mohr medium and cultured in a climatic chamber in the conditions described above.

2.2 Procedure

Fragments of *C. foliacea* thalli were stored dry. Before use, these were abundantly rinsed in tap water and rapidly surface sterilized (10 seconds) in 1% sodium hypochloride. They were then inoculated in the in ratio of 0.5 g *Cladonia* thallus to each Petri dish containing 20 ml medium (water or Mohr) with 200 µl spore suspension (see above) or 20 shoot fragments.

2.3 Observations and Data Analysis

(i). Cultures are conducted in triplicate and maintained in a growth chamber at 13° (night) to 20° C (day), 70% humidity with 16 h light (20-50 Wm^{-2}) / 8 h dark cycle, observed after 3, 5, 7, 14, 21 days up to 2, 3 months and images recorded with a microscope, a stereomicroscope and/or an inverted light microscope equipped with a photocamera.

(ii). Spore germination (%) and protonemal growth are evaluated in aseptic cultures by examining under an inverted microscope about 300 spores per replicate.

(iii). Number of filaments per axillary node (or leaf for *P. squarrosa* or germinating propagules for *L. cruciata*) are recorded. Protonemal length is measured on 30 sporelings or shoot derived protonemal filaments per replicate. Significance of differences is evaluated by Anova or Student's t test as appropriate.

3. GUIDED ISOLATION OF POTENTIAL ALLELOCHEMICALS

Experiment: The isolation and identification of bioactive molecules is a major goal of chemical ecology and can be achieved by guided bioassays on suitable sensitive systems. Here a bioassay based on the growth and morphogenesis of protonemal system of bryophytes grown *in vitro* is described to test potential allelochemicals from lichen *Cladonia foliacea*.

3.1   Materials Required

Plant material and protonemal systems from spores and/or shoots are same as described in experiment 1.

Test fractions/extracts: Dried and powdered thalli (20 g) of *C. foliacea* were continuously and progressively Soxhlet extracted for 8 hour as per Giordano et al. (1997), obtaining 8 fractions whose purification and identification of active constituents were guided by bioassays on spore germination and protonemal growth. Polar extracts were diluted with two drops of DMSO (dimethyl sulfoxide) and then dissolved in Mohr medium. The other extracts were directly dissolved in Mohr medium to obtain a dry weight lichen thallus/Mohr medium ratio of 0.1 mg/ml. The obtained solutions were biologically sterilized through millipore filters (0.45 μm diameter).

3.2   Procedure

The cultures were obtained and maintained in the same conditions described above and observed with the same procedure at the same time intervals. The different morphogenetic effects induced by active fractions are evaluated by the following parameters: (i) number of cells in the main filament, (ii) kind of filaments (caulonema, chloronema, rhizonema) (Figure 2), (iii) order of secondary branches, (iv) number of cells without ramification from the apex, (v) number of cells in the I and II order filaments (Table 1). Significance of differences is evaluated by Anova or Student's t test as appropriate.

3.3   Disadvantages

In vitro cultures of bryophytes, especially when obtained from wild material, may be contaminated by fungi and/or bacteria. Therefore, we suggest starting the cultures from mature sporophytes that can be more conveniently surface sterilized.

4. CONCLUSIONS

The whole *Cladonia* thalli and its derived fraction delay the spore germination and regeneration of new filaments and also inhibit their growth. Some alterations at cell level are : the granular appearance and the presence of microvesicles in cytoplasm and the alterations in the shape of chloroplast, which become spherical and bulged. In cultures with whole *Cladonia* thalli, moss protonemal filaments are significantly shorter than

Table 1 Experiment 2. Guided isolation and identification of potential allelochemicals from *Cladonia foliacea*. O, fraction O; P, fraction P; UA, usnic acid; ca, caulonema; ch, chloronema; sb, side branches. Protonemal system from shoots after 21 days in culture tested against three fractions from *C. foliacea* thallus and in control Mohr medium. The observations represent means calculated from about 300 protonemata for each species. *P. squarrosa* is the species least affected, while *T. flavovirens* is the most affected by lichen fractions.

Parameters	P. squarrosa				T. flavovirens			
	Mohr	O	P	UA	Mohr	O	P	UA
Number of cells of the main filament	16.2	18.7	12.2	12.3	12.2	27.7	0	0
Kind of filaments	ca, ch	ca, ch	ca, ch	ca, ch	ca, ch	ca, ch	-	-
Order of side branches	I, II	I, II	I, II	I, II	I, II	I, II, III	-	-
Number of cells from apex for emergence of side branches	5.3	3.8	2.2	2.4	1.3	1.8	-	-
Number of cells of I order side branches	9.3	12.8	6.8	6.2	4.3	5.8	-	-
Number of cells of II order side branches	5.2	4.7	5.6	4.3	2.2	2.7	-	-

controls (P < 0.05), but cell size seems constant, suggesting that *Cladonia* reduces the speed of cell division. All the same, since rhizoids produced by germinating gemmae of *L. cruciata* are unicellular, the inhibiting effect in this species acts also on cell elongation (Figure 3).

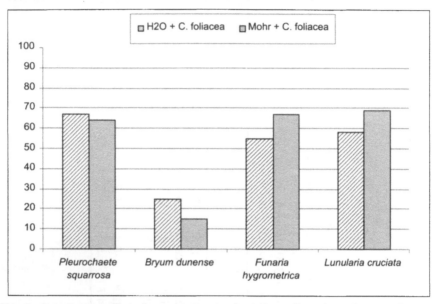

Fig. 3   Experiment 1: Test of co-existence *in vitro* between bryophyte shoots and lichen thalli. The histograms indicate the percentage of newly formed filaments in cultures obtained by regeneration in the presence of *Cladonia* thalli after the first week, compared to control (100%) in water or Mohr. The results represent the average calculated from about 200 shoots for each species. SD was not more than ± 13%. (modified from Giordano S., et. al : *Cryptogamie, Bryol.* 20(1): 35-41, 1999).

Among the fractions obtained during the isolation/purification of allelochemicals from *Cladonia foliacea*, two had inhibiting effects and one was stimulating for spore germination and protonemal growth. Of the two fractions showing inhibiting properties, one was found to be usnic acid (UA), the other (fraction P) consisted > 95% oxidisable aldehyde, hence not identified. The stimulating fraction (O) was a mixture of different alditols viz., mannitol and arabitol, molecules commonly used as a carbon source in many culture media (Figure 4).

Spore germination and protonemal growth and morphogenesis are therefore, useful systems to test *in vitro*, with good reproducibility, potential allelochemicals both through direct co-existence test *in vitro* or using the described bioassays to monitor/guide isolation, purification, characterization of chemical structure of bioactive compounds.

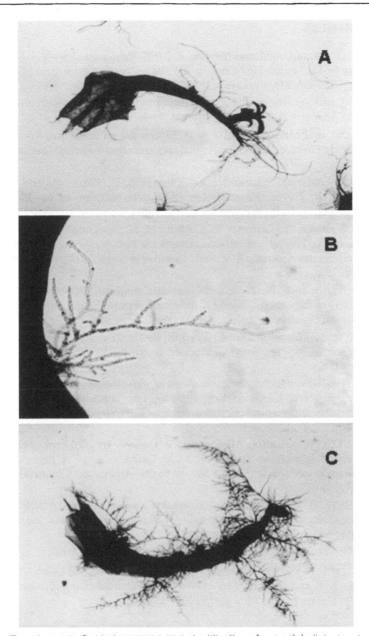

Fig. 4 Experiment 2: Guided isolation and identification of potential allelochemicals from *Cladonia foliacea*. A, Regenerating protonema from *Pleurochaete squarrosa* leaves after 14 days in control sample; B, regenerating protonema from *P. squarrosa* after 14 days in the inhibiting fraction P from lichen *Cladonia foliacea*;. C, regenerating protonema from *P. squarrosa* after 14 days in the stimulating fraction O from lichen *Cladonia foliacea*. Light microscope (a and c x 25, b x 50).

5. REFERENCES

Duckett J. G., Schmid, A.M. and Ligrone, R. (1998). Protonemal morphogenesis. In: *Bryology for the Twenty-First Century.* J.W. Bates, N.W. Ashton and J.G. Duckett (eds). pp 223-246. Maney, Leeds, U.K.

During, H.J. and Van Tooren, B.F. (1990). Bryophyte interactions with other plants. *Botanical Journal of the Linnean Society* **104**: 79-98.

Gardner, C.R. and Mueller D.M.J. (1981). Factors affecting the toxicity of several lichen acids: effect of pH and lichen acid concentration. *American Journal of Botany* **68**: 87-95.

Giordano, S., Alfano, F., Esposito, A., Spagnuolo, V., Basile, A. and Castaldo Cobianchi, R. (1996). Regeneration from detached leaves of *Pleurochaete squarrosa* (Brid.) Lindb. in culture and in the wild. *Journal of Bryology* **19**: 219-227.

Giordano, S., Basile, A., Lanzetta, R., Corsaro, M. M., Spagnuolo, V. and Castaldo Cobianchi, R. (1997). Potential allelochemicals from the lichen *Cladonia foliacea* and their in vitro effects on the development of mosses. *Allelopathy Journal* **4** : 89-100.

Giordano, S., Alfano, F., Basile, A. and Castaldo Cobianchi, R. (1999). Toxic effect of the thallus of the lichen *Cladonia foliacea* on the growth and morphogenesis of bryophytes. *Cryptogamie Bryology* **20**(1): 35-41.

Knoop, B. (1984). Development in Bryophytes. In: *The Experimental Biology of Bryophytes* , A.F. Dyer and J.G. Duckett (eds), pp 147-151. Academic Press, London, UK.

Krupa, J. (1964). Studies on the physiology of germination of spores of *Funaria hygrometrica* (Sibth.). I. The influence of light on germination with respect to water balance and respiratory processes. *Acta Societatis Botanicorum Poloniae* **33**: 177-192.

Lawrey, J. D. (1995). Lichen Allelopathy: A Review. In: *Allelopathy Organisms, Processes and Applications* . Inderjit, K. M. M. Dakshini, and F.A. Einhelling (eds), ACS Symposium Series. No. **582** : 26-38 American Chemical Society, Washington DC., USA.

Nichols, H. W. (1973). Growth media - fresh water. In: *Handbook of Phycological Methods, Culture Methods and Growth Measurement*. J. R. Stein (ed), pp 7-24. Cambridge University Press, Cambridge, UK.

Rice, E.L. (1979). Allelopathy-an update. *Botanical Review* **45**: 15-109.

Rice, E.L. (1984). *Allelopathy*. 2nd ed. Academic Press, New York. USA .

Schumaker, K. S. and Dietrich, M. A. (1997). Programmed changes in form during moss development. *The Plant Cell* **9**: 1099-1107.

Section 2

New Methods of Microscopy in Cellular Diagnostics

5

Chapter

Microscopic Methods to Study Morpho-cytological Events during the Seed Germination

G. Aliotta[1], *G. Cafiero*[2] *and M. Petriccione*[1]

1. INTRODUCTION

A common test for the allelopathic potential of a plant *in vitro* is to determine the effects of its aqueous extract on the seed germination and subsequent radicle growth. The seed is an ideal tool because it is a dispersal unit of life, consisting of three genetically different parts: seed coat, endosperm and an embryo able to sprout out as a new plant, if exposed to the right conditions (Evenari, 1980). This methodical paper deals with the morphological and cytological responses of two different seeds viz., radish (*Raphanus sativus* L. 'Saxa') and purslane (*Portulaca oleracea* L.), sown with rue (*Ruta graveolens* L.) extract. We have made efforts to simplify the basic microscopic techniques for a beginner.

About 75% biologically active plant-derived compounds presently used worldwide, have been discovered through follow-up research to verify the authenticity of folk medicinal plants (Farnsworth, 1990). Rue (*Ruta graveolens* L., Rutaceae) is an evergreen plant native of Southern Europe, with bluish-green leaves that emit a powerful odour and have a bitter taste. The plant is cited in the ancient herbal alchemy. Rue has been regarded from the earliest time as successful in warding off contagion and preventing the

[1] Dipartimento di Scienze della Vita, Seconda Università di Napoli Via Vivaldi 43, I-81100 Caserta, Italy.
[2] Centro Interdipartimentale di Servizio per la Microscopia Elettronica, Università di Napoli Federico II Via Foria 223, I-80139 Napoli, Italy, E-mail: giovanni.aliotta@unina2.it

attacks of fleas and other noxious insects. The name rue derives from the Greek 'reuo' (= to set free), because the plant is free from various diseases. It produces erythema and pustular eruptions on the human skin. Many remedies containing rue as well as its abortive properties were mentioned by Pliny the Elder (23-79 AD) in his *Naturalis Historia* (XX, 143). Piperno, a Neapolitan physician, in 1625, recommended rue as a treatment for epilepsy and vertigo. Today, its aerial parts are eaten in Italian salads and are said to preserve eyesight. Rue is currently mentioned in the pharmacopoeias of 28 countries, as stimulating, antispasmodic, diuretic and emmenagogue. Moreover, fresh and dried leaves are used to preserve and to flavour beverages and foods such as liquor (grappa) and wine, cheese and meat (Aliotta et al., 1995). Secondary chemical constituents and potential allelochemicals of rue are listed in Table 1. The term allelochemical coined by Whittaker and Feeney (1971) recognized that many naturally produced substances usually called secondary metabolites have the ability to affect the growth, health, population biology or behaviour of another species.

Besides its ethnobotany, we have selected rue for allelopathic studies because of the following reasons:

a) Our previous studies established coumarins as potent allelochemicals (Aliotta et al., 1993);
b) Rue and basil never grow together or near each other (Grieve, 1967);
c) The presence of large amount of coumarins on the leaf surface of rue and their easy extraction through leaching makes the plant extract an ideal cheap tool for allelopathic studies (Zobel and Brown, 1988).

2. PREPARATION OF RUE EXTRACT

Fresh leaves of rue were collected from the plants grown in the Botanical Garden, Naples. Two hundreds grams (fresh weight) of leaves were extracted directly by dipping them for 10 min in 1 litre of hot water (95°C). The resulting infusion was subsequently used for the germination experiments.

2.1 Radish (*Raphanus sativus*)

2.1.1 Morphological and cytological responses.

Rue infusion was tested for its allelopathic activity *in vitro* on water uptake and germination of radish seeds. Seeds of radish (*Raphanus sativus* L.) 'Saxa', collected during 2003, were purchased from Improta Co., Naples.

2.1.1.1 Water uptake by radish seeds

Experiment 1. *To determine the water uptake of radish seeds*

Table 1 Secondary chemical constituents and potential alleochemicals of *Ruta graveolens* (modified from Murray et al., 1982). Allelochemicals are marked with an asterisk.

Compound	Plant part	Compound	Plant part
Coumarins		***Alkaloids***	
Bergapten*	Stems, leaves, cell cultures	Arborinine	
(-)-Byakangelicin	Roots	g-Fagarine	
Chalepensin	Roots	Graveoline	
Coumarin*	Leaves	Graveolinine	
Daphnoretin	Aerial parts	Kokusaginine	
Daphnoretin methyl ether	Roots	6-Methoxy-dictamine	
Daphnorin		Rutacridone	
Gravelliferone	Roots	Skimmianine	
Gravelliferone methyl ether	Roots	*Flavonoids*	
Herniarin	Cell cultures	Quercetin*	Leaves
Isoimperatorin	Roots and stems	Rutin*	Leaves
Isopimpinellin	Cell cultures	*Ketones*	
Isorutarin	Cell cultures, roots	Methyl-nonyl-ketone	Aerial parts
Marmesin	Cell cultures, roots	Methyl-heptyl-ketone	Aerial parts
Marmesinin	Roots	*Organic acids*	
8-Methoxy-gravelliferone		Anisic acid	
Pangeline		Caprinic acid	
Psoralen*	Cell cultures, roots, stems, leaves	Caprylic acid	
Rutacultin	Roots	Oenanthylic acid	
Rutamarin	Aerial parts, roots, stems	Plagonic acid	
Rutamarin alcohol	Roots	Salicylic acid*	
Rutaretin	Leaves (?)	*Terpenoids*	
Rutarin	Aerial parts, roots	Cineole*	Aerial parts
Scopoletin	Cell cultures	Limonene*	Aerial parts
Suberenon	Roots	Pinene*	Aerial parts
Umbelliferone*	Cell cultures, roots	Guaiacol	Aerial parts
Xanthotoxin*	Cell cultures, stems, leaves		
Xanthyletin	Roots		

*Allelopathic compound

Procedure: Ten radish seeds (80±5 mg) were sown in each Petri dish (9-cm diameter) on three layers of filter paper (Whatman No. 3) previously moistened either with 7 ml of distilled water (control) or 7 ml of rue infusion (treated). Eighteen hours after sowing, control and rue treated radish seeds were differently moistened, having weight increases of 110 and 80 % respectively. The water uptake was evaluated as the difference in weight between moist and dry seeds (% of initial weight)(Fig.1.).

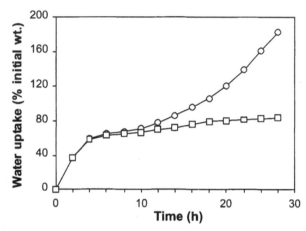

Fig. 1 Water uptake by radish seeds in presence (o) and absence (□) of rue extract (on a percent of initial weight of seeds).

2.1.1.2 Germination of radish seeds

Experiment 1. *To determine the germination of radish seeds*

Procedure: Germination conditions were 25±1°C under continuous fluorescent light of 25 mE m^{-2} sec^{-1}. Seed germination was monitored by observing the seeds directly in the Petri dishes with a stereomicroscope. They were considered germinated, when the radicle had protruded through the seed coat. Seeds sampled at different times after the beginning of imbibition were used for microscopy studies.

Figure 2 reports the time dependent germination pattern of radish seeds in the presence and absence of rue extract. Although the germination curves have the same general shape, important differences between control and treated seeds are evident. In fact, the rue extract induces a delay in the onset and a decrease in the rate of germination.

2.1.1.3 Anatomical and ultrastructural changes in radish seeds

By means of stereo, light and electron microscopy, we have studied the anatomical and ultrastructural aspects of hilum-micropylar region of the radish seed, where the radicle protrudes.

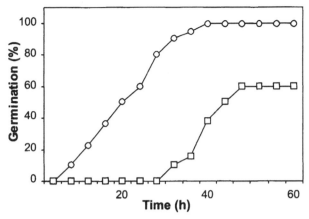

Fig. 2 Time course of germination of radish seeds in presence (○) and absence (□) of rue extract. Germination is expressed as percentage of germinated seeds.

I. Stereomicroscopy of radish seeds: The studies were carried out after 18 h in the presence and absence of rue extract, when most of the seeds soaked in water were germinating. Treated seeds were dormant, and the different uptake of water into these seeds was evident.

Experiment 3. *Observations under stereomicroscope*
The image of radish seed under stereomicroscope is shown in Fig. 3. According to Vaughan and Whitehouse (1971), radish seed is oval (3 ´ 2 mm) with a light brown to orange, reticulate surface. The seed coat of the mature seed has 3 layers of cells viz., (i) Epidermis, (formed by compressed cells); (ii) the palisade layer (presents cells more or less isodiametrical or radially elongated) and (iii) inner pigmented parenchyma layer (it is one cell thick). The endosperm persists as a well-formed aleurone layer intimately associated with the seed coat.The hyalin layer covers the embryo (Figure 4). The embryo is folded with cotyledons against radicle (Aliotta et al., 1994).

II. Scanning electron microscopy of radish seeds: For scanning electron microscopy (SEM) observations, seed coats and endosperms were fixed in 3% glutaraldehyde in 0.065 M phosphate buffer (pH 7.4) for 2 h at

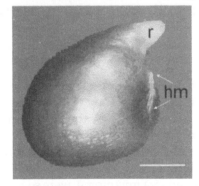

Fig. 3 Stereomicrograph of a germinating seed of radish showing the hilum micropylar region (hm) and the radicle (r), 18 h after sowing. Bar = 1mm

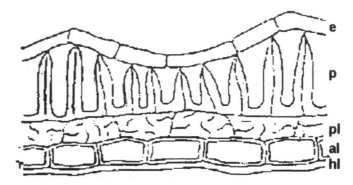

Fig. 4 Graphic showing the structural features of a seed coat and endosperm of a radish seed, according to Vaughan and Whitehouse (1971). e, epidermis; p, palisade; pl, pigment layer; al, aleurone layer; hl, hyalin layer.

room temperature. The specimens were then placed into 2% OsO_4 in 0.1 M phosphate buffer (pH 6.8) at 4°C before being dehydrated with ethanol and propylene oxide, critical-point dried and finally coated with carbon and gold in a sputter-coater. These specimens were observed at 20kV with a Cambridge 250 Mark3 scanning electron microscope.

Experiment 4. *Observation under electron microscope.*

Figure 5 shows a comparison between control and rue treated radish seeds by stereo and SEM microscopy. Moistened seeds were excised (cut in half along the two orthogonal planes of their major axis) and the embryos were removed. These sections were directly observed with a Steromicroscope Wild M3Z. As shown in Figs. 5 A and B, the section of hilum-micropylar region control of seed (A) presents two bands having different colours: the external band is black and the inner is grey. This latter does not appear in the treated seed (B).

SEM micrographs of the hilum-micropylar region showed that the hylum was more evident in the control than in the treated seed (Fig. 5 C, D).

III. Transmission electron microscopy of radish seeds: Transmission electron microscopy (TEM) of radish seeds was done as listed below: For TEM preparations, the specimens after fixation and dehydration, were embedded in Epon 812 resin (Luft, 1961). Thick sections (ca. 1mm each) were stained with 0.1% toluidine blue and observed with a Zeiss light photomicroscope. Thin sections, obtained with a diamond knife on a Supernova microtome, were sequentially stained at room temperature with 2% uranyle acetate (aqueous) for 5 min and by lead citrate for 10 min (Reynolds, 1963). Ultrastructural studies were made using a Philips CM12 transmission electrone microscope (TEM) operated at 80 KV.

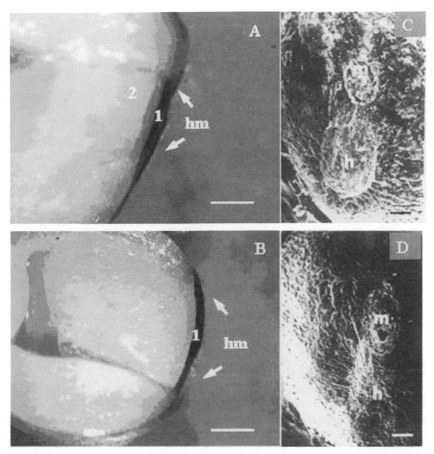

Fig. 5 An overview of control and treated radish seeds. A,B Stereomicrographs of radish seeds, moistened (A) and rue-treated (B), 18 h after sowing. Cut seeds were excised along the two orthogonal planes of their major axis. The hilum-micropylar region shows two bands in (A) and one in (B). Bars = 0.5 mm. C,D. SEM micrographs of the hilum-micropylar region of radish; 18 hr after sowing the hilum is more evident in water moistened seed (C), than in rue-treated seed (D). m, micropyle and h, hilum. Bars = 0.5 mm

Experiment 5. *Observation under transmission electron microscope*

We compared the TEM ultrastructure of the seed coat and endosperm of control and rue-treated seeds The palisade layer of treated seed appears thicker than in the control (Figs 6A and 7A), while comparison between aleuronic cells of the control and treated cells (Figs. 6B and 7B), reveals that the cells of the control are healthy with some evident organelles such as the nucleus and the rough endoplasmic reticulum and other structures, the plastid, the plasmodesmata, conspicous constrictions, protein bodies and

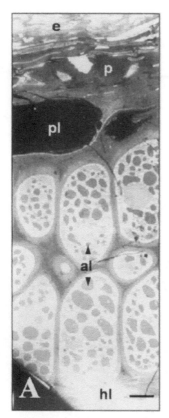

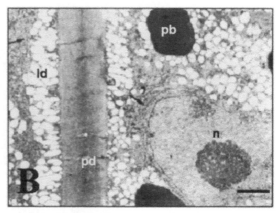

Fig. 6   TEM micrographs of seed coat and aleurone cells of radish control seed 18 h after sowing in water; e, epidermis; pl, pigment layer; al, aleurone layer. (A) Bar = 30 μm; Particulars of the aleurone cell showing some organelles: nucleus (n), plasmodesmata (pd), protein bodies (pb) and lipid droplets (ld). (B, C). Bar = 5 mm.

lipid droplets. The differences observed between moistened and rue extract-treated radish seeds represent useful signals to establish whether the water uptake of the seed will culminate in radicle emergence. These findings suggest that in radish seeds, rue extract in the presence of light provides, either directly or indirectly, a critical signal that inhibits water uptake into the seed, switches off the swelling of the seed coat epidermis as well as of the cell of the endosperm aleurone layer and the related process of cell elongation in the embryo. Thus, it is possible that germination in the radish results from a combination of lowered resistance of the swollen seed coat, endosperm, and of the related elongation on embryo cells. These effects due to the active coumarins are prevented by a preliminary removal of seed coats or by darkness (not shown).

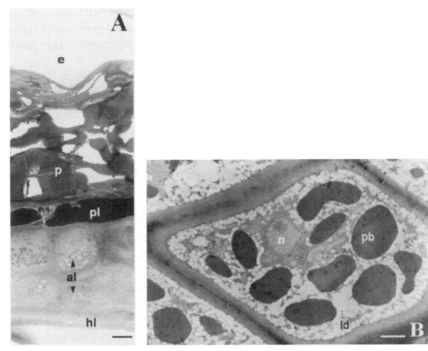

Fig. 7   TEM micrographs of seed coat and aleurone cells of radish treated seed 18 h after sowing in presence of rue extract. e, epidermis; pl, pigment layer; al, aleurone layer (A). Bar = 30 μm. Particulars of aleurone cell showing different profile respect to those of control. ld, lipid droplets; pb, polar bodies; n, nucleus. (B). Bar = 5 mm.

2.2   Purslane (*Potulaca oleracea*)

Based on the results from radish seeds, we also studied the influence of the rue extract on germination of purslane. Purslane is a major weed worldwide of 45 crops in 81 countries. The ploughable layer of the soil cropped with maize contains about 220,000 purslane seed per m². Purslane fruit is a capsule with a pyxidium and a caliptra containing 50-70 black seeds (Holm et al., 1977).

2.2.1   Germination of purslane seeds

Procedure: Purslane seeds were collected from crop fields near Naples. Five hundred seeds were sown in 10 Petri dishes (∅ =90 mm), containing 5 layers of Whatman filter paper impregnated with 7 ml of water (control) or 7 ml rue infusion/chromatographic fractions or isolated compounds as per treatment. Thereafter, daily 3 ml water was added to each Petri plate. Germination conditions were 30±1°C with a continuous light of 25μE/m²/

sec. Seeds germination process was observed directly in Petri dishes with a stereomicroscope at 8 h intervals. A seed was considered germinated when the protrusion of the radicle became evident. Each experiment was repeated thrice.

2.2.2 Scanning electron microscopy of purslane seeds

Anatomical and ultrastructural aspects of *in vitro* germination of the purslane seeds may be studied using stereo and SEM microscopy (Figures 8 and 10).

Procedure: A dry quiescent purslane seed and the successive changes when it is moistened have been reported here.

Experiment 1. *Morphology of purslane seeds treated with rue extract.*
The outer integument (testa) of the seed coat is formed by dead cells sculputered on the surface with stellulae (Figure 8A). The peripheral face of the testa presents an opening; the micropyle and a residue of the funiculus of the placenta which functions as elaiosome. Surprisingly, the first structure that protrudes from the micropyle of the moistened seed, is not the radicle, as in seeds from most species, but the endosperm of the seed coat (Figure 8B). When the germination proceeds further, the radicle breaks the endosperm and protrudes (Figure 8C).

2.2.3 Purslane seed germination

Procedure: Same Procedure as per Section 2.1.1. was used.

Experiment 2. *Time course of endosperm protrusion and germination with and without rue infusion*
Rue infusion delayed and decreased the germination of purslane seeds (Fig. 9). In control after 30 h moistening, only 25-30% seeds had protruded endosperm or germinated and the maximum endosperm and radicle protrusion occurred between 40-60 per h and all seeds germinated at 70 per h In rue infusion treated seeds, most remained dormant and no morphological changes were noted up to 30 h. At 70 and 110 h 20 and 40% seeds protruded endosperm and only 5% and 20% germinated, respectively. Both types of seeds i.e. dormant or with protruded endosperm that failed to germinate up to 120 h in the presence of the rue infusion, germinated within 3-4 h of being transferred to control conditions. On the contrary, when the controlled seeds with protruded endosperm were transferred to Petri dishes containing rue infusion, germination was blocked and the radicles failed to emerge. This reversible action of rue infusion, indicates the important role of

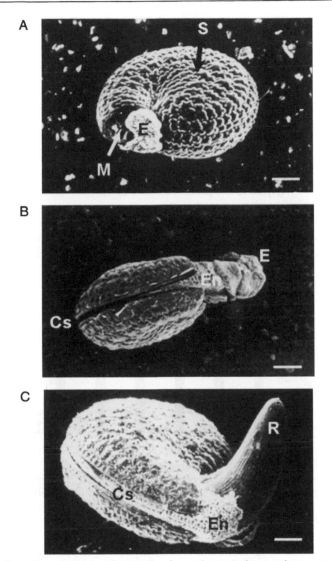

Fig. 8 Scanning electron micrograph of a quiescent dry purslane seed (A) and its successive modifications after 25 and 30 h respectively (B,C). Cs = Circumcision of lateral face. E = Elaiosome. En = Endosperm. M = Micropyle. R = Radicle. S = Stellula. Bar = 0.1 mm.

purslane endosperm in regulating its germinaion. As already reported for radish seeds, the outermost living cells of the seeds micropylar region may be responsible for the *primum movens* of the germination process and represent the primary target for rue extract.

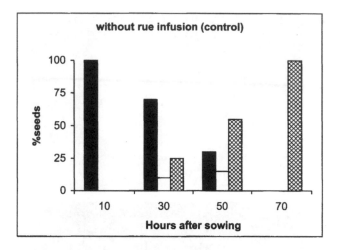

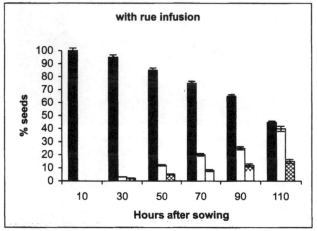

Fig. 9 Time course of endosperm protrusion and germination of purslane seeds without rue infusion (control) and with rue infusion. Histograms represent the mean ± SD of percentage of dormant (■), protruded endosperm (⊠) and germinated seeds □.

Experiment 3. *Scanning electron microscopy of developed pursalame seedlings.*
Once the radicle of purslane protrudes in the absence of rue, it rapidly completes the early stages of primary structure development viz., root hairs and secondary roots (Figure 10A). While in the presence of rue infusion, if endosperm regulation fails, the radicle protrudes but is unable to grow and become irreversibly damaged, however, the hypocotyl and cotyledons show normal growth (Figure 10B) (Aliotta et al., 1996).

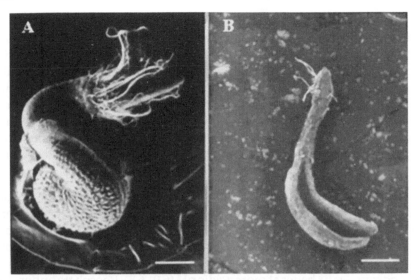

Fig. 10 Scanning electron micrographs of a germinating seed of purslane in absence (A) and in presence of rue infusion (B).

3. PREPARATION OF TEST SEEDS FOR ELECTRONIC MICROSCOPY

Important steps for preparation of radish and purslane seeds for microscopy are listed below :

I. Fixation: Fixation is an attempt to quickly arrest biological activity and to stabilize cellular components with minimal distortion of conformational and spatial relationships between cellular constituents. The purpose of fixation is to preserve tissue structure with minimal alteration during dehydration, embedding, cutting, staining and viewing the microscope. The most common reason for poor fixation is large specimen size. For example, glutaraldehyde, the fixative most often used in electron microscopy, penetrates to a depth of less than 1 mm. To minimize autolytic changes, slices or ribbon of tissue 0.5 mm thick should be placed in fixative promptly (Slayter and Slayter, 1993).

   Chemical fixation for transmission electron microscopy prepares cells for the preservation of damage due to subsequent washing with aqueous solvents, dehydration with organic solvents such as ethanol or acetone, embedding in plastic resins, polymerization of the resins by heat, exothermic catalysts, or ultraviolet radiation, and imaging with high-energy electron beams in an electron microscope.

Materials prepared for scanning electron microscopy are not embedded with resins, but while in solvents they are subjected to high pressures during critical point drying. An ideal fixative would transform the viscous colloidal protoplasm of a cell into cross-linked and stabilized cellular components. The spatial relationship between all organelles and cellular structures are not altered, the cellular components are not solubilized, and the biological activity of complex proteins like antigens and enzymes remain undiminished (Dykstra, 1993). Moistened seeds were excised (cut in half along the two orthogonal planes of their major axis), the embryos were removed and the seed coats and endosperms were fixed in 3% glutaraldehyde.

II. Buffering: The buffering is done to stabilize the pH of the tissue, somewhere near physiological levels during fixation and to maintain nearisotonic conditions. For structural studies, most cells fix well within pH range from 7.0 to 7.4. Certain highly hydrated tissues fix better at more alkaline pH (8.0 -8.4), whereas plant cells, nuclear material and the fibrils of mitotic spindles fix better at more acidic pH (6.0 - 6.8). Phosphate buffers are typically used at physiological pH. Phosphate buffer of 0.065 M (pH 7.4) was used. They are specifically avoided in situations where the negatively charged phosphate groups could be expected to interact with positively charged ions in incubation media, causing precipitates to form. There are three major phosphate buffers used in most laboratories: Sorenson's phosphate buffer formulated from sodium and potassium phosphate salts and Millonig's sodium phosphate buffer (Glauert, 1975; Millonig, 1964).

III. Resin Embedding: Spurr' resin was originally designed for plants with cell walls that were hard to infiltrate with higher-viscosity epoxide resins. It requires only a relatively brief infiltration schedule for most materials and thus can shorten overall specimen processing time. The embedding resin is compatible with acetone, ethanol and other commonly used solvents. The resin is completely mixed with ethanol, though blocks polymerized following an ethanol only dehydration series often have a slightly tacky surface not found when acetone is used as a transitional solvent. They also may be slightly more brittle. Due to the low viscosity of Spurr' resin, it is not necessary to rotate or agitate specimens in the resin during infiltration. With proper handling, Spurr' resin provides quick infiltration of difficult samples and can be readily trimmed and sectioned, yielding excellent results (Spurr, 1969).The specimens were then placed into 2% OsO_4 in 0.1 M phosphate buffer (pH 6.8) at 4OC before being dehydrated with ethanol and propylene oxide, and embedded in Spurr's resin.

IV. Cutting Sections: Ultramicrotomes are designed to cut ultrathin sections, semithin sections and ultrathin frozen sections if suitably equipped. An ultramicrotome with either a thermal or a mechanical advance mechanism will work well for sectioning plastic-embedded samples. If ultrathin frozen sections are needed for the research objectives anticipated, mechanical advance units offer some advantages in terms of consistency of section thickness. (Fig. 13). Semithin sections up to several square millimeters in size and 0.25 – 0.5 mm thick can be cut easily from epoxy or acrylic resin blocks containing fixed samples with glass knives (Fig. 13A) and examined by light microscopy. Ultrathin sections approximately 0.5 mm^2 in size and 80 – 90 nm thick can be cut with either glass (Fig. 13 B,C) or diamond knives for examination by TEM (Fig. 13 D,E,F). Thick sections (ca. 1 mm each) were stained with 0. 1% toluidine blue and observed with a Zeiss light photomicroscope. Thin sections were cut with a diamond knife on Supernova microtome.

V. Post Fixation: The fixation of biological samples for scanning electron microscopy (SEM) involves all the same principles for preserving specimen structural integrity. Osmium tetroxide postfixation can sometimes be omitted, though samples that have charging problems leading to image distortion can frequently benefit from osmication. Subsequent dehydration in an ethanol or acetone series followed by critical point drying, freeze-drying, or drying with one of the chemical techniques produces a sample that can be introduced into the high vacuum system of the SEM. The dried sample is then attached to support stubs with a variety of materials (colloidal silver, colloidal carbon, double sided tape, or conductive carbon tape, and other) prior to coating with precious metals such as gold-palladium to ensure the electrical conductivity of the specimen surface. Normally, biological SEM samples are examined in the secondary electron imaging mode (Dykstra, 1993). Seed coats and endosperms of control and treated seeds, after fixation and were dehydrated with ethanol.

VI. Critical Print Drying: This technique is used for samples of soft tissue that are hydrated. If the sample is already dry (bone, prosthetic bone implants, air-dried field samples of pollen, spores, etc.), critical point drying is often inappropriate. In such cases, the dried sample can be immediately mounted on SEM stubs, sputter coated and examined with the SEM. A hydrated biological sample should be fixed and dehydrated as described for routine SEM preparation. The dehydration fluid (ethanol) must be removed from the specimen prior to sputter coating and viewing with an SEM. Samples were critical point dried. If the sample is allowed to air-dry, pressures up to 2000 1b/in^2 can be generated, resulting in major distortion of specimen surfaces. Critical point drying is based on the concept that at a certain temperature and pressure, the vapour and liquid phases of carbon dioxide become indistinguishable.

Liquid CO_2 from a siphon-tube tank is introduced into the chamber and used to replace 100% of the ethanol in the specimen. After the ethanol has been totally replaced by the CO_2, the critical point drying chamber (CPD) is brought above the critical point. The temperature is kept above the critical point, while the gaseous CO_2 is vented from the chamber. The process is finished when the CPD is returned to atmospheric pressure. After critical point drying, the specimen should be totally dry and it is ready to be introduced to the vacuum system of the sputter coater and SEM (Dykstra, 1993; Hayat, 1978).

VII. Coating: Sputter coating is a technique for deposing a metal coating on specimen surfaces to be examined with a SEM. The metal is deposited non directionally, allowing even coating surfaces of the specimen, regardless of topography, unlike the directional coating process of vacuum evaporation. Less specimen heating is developed during sputter coating than with vacuum evaporation. The sample was finally coated with carbon and gold in a sputter-coater.

Although this chapter may not provide complete and final synthesis of the subject, our hope is that the information presented can stimulate young researchers to expand both the depth and breath of microscopy in allelopathy.

4. REFERENCES

Aliotta, G., Cafiero, G., De Feo, V., Palumbo, A.D. and Strumia, S. (1996). Infusion of rue for control of purslane weed: biological and chemical aspects. *Allelopathy Journal* 3: 207-216.

Aliotta, G., Cafiero, G., De Feo, V. and Sacchi, R. (1994). Potential allelochemicals from *Ruta graveolens* L. and their action on radish seeds. *Journal of Chemical Ecology* 20 : 2761-2775.

Aliotta, G., Cafiero G., Fiorentino, A. and Strumia, S. (1993). Inhibition of radish germination and root growth by coumarin and phenylpropanoids. *Journal of Chemical Ecology* 19: 175-183.

Aliotta, G., Cafiero, G., Oliva, A. and Pinto, E. (1995). Ethnobotany of Rue (*Ruta graveolens* L.) : an overview. *Delpinoa* 37-38: 63-72.

Dykstra, M.J. (1993). *A Manual of Applied Techniques for Biological Electron Microscopy.* Plenum Press. New York,USA.

Evenari, M. (1980). The history of germination research and the lesson it contains for today. *Israel Botany* 29:4.

Farnsworth, N.R. (1990). The role of ethnopharmacology and the search for new drugs. In: *Bioactive Compounds from Plants, .,* J. Chadwick and J. Marsh (eds) pp.2-21. John Wiley & Sons, Chicester U.K.

Glauert, A. M. (1975). *Fixation, Dehydration and Embedding of Biological Specimens.* North-Holland, Amsterdam,the Netherlands.

Grieve, M. (1967). *A Modern Herbal.* Hafner Publishing Co. London,UK.

Hayat, M.A. (1978). *Principles and Techniques of Scanning Electron Microscopy : Biological Applications.* VNR Company, New York,USA.

Holm, L. G., Plucknett, D. L., Pancho, J. V. and Herberger, J. P. (1977). *The World's Worst Weeds.* The University Press of Hawaii, Honolulu,USA.

Inderjit (2002). Multifaceted approach to study allelochemicals in an ecosystem. In: *Allelopathy From Molecules to Ecosystems,* , M. J. Reigosa and N. Pedrol (eds) : Science Publishers, Inc. Enfield Plymouth, USA.

Millonig, G. (1964). Study on the factors which influence preservation of fine structure. In: *Symposium on Electron Microscopy* P. Buffa (ed), p. 347. Consiglio Nazionale delle Ricerche, Rome, Italy.

Murray, R.D.H., Mendez, J. and Brown, S.A. (1982). *The Natural Coumarins: Occurrence, Chemistry and Biochemistry.* John Wiley and Sons, Chichester, U.K.

Reynolds, E.S. (1963). The use of lead citrate at high pH as an electron opaque stain in electron microscopy. *Journal of Cell Biology* **17**: 208-212.

Slayter, E.M. and Slayter, H.S. (1993). *Light and Electron Microscopy.* Cambridge University Press, Cambridge,UK.

Spurr, A.R. (1969). A low-viscosity epoxy resin embedding medium for electron microscopy. *Journal of Ultrastructural Research* **26**: 31.

Vaughan, J.G. and Whitehouse, J.M. (1971). Seed structure and the taxonomy of the Cruciferae. *Botanical Journal of Linnean Society* **64**: 383-409.

Whittaker, R. H. and Feeney, P.P. (1971). Allelochemics: Chemical interaction between species. *Science* **171**: 210-218.

Zobel, A.M. and Brown, S.A. (1988). Determinations of furocoumarins on the leaf surface of *Ruta graveolens* with an improved extraction technique. *Journal of Natural Products* **51** : 941-946.

Chapter

Optical Coherence Tomography and Optical Coherence Microscopy to Monitor Water Absorption

V.V. Sapozhnikova, I.S. Kutis, S.D. Kutis, R.V. Kuranov,
G.V. Gelikonov, D.Y. Shabanov and V.A. Kamensky.

1. INTRODUCTION

Many optical non-invasive methods for monitoring of plant tissue are currently being developed. Recently, new non-destructive methods providing structural maps i.e., images of tissues of plants and animals, have been reported. Confocal, two- and multiphoton microscopy are very promising methods for plant physiology research (Haseloff, 1999). Some of the non-invasive methods vizualization, optical coherence tomography (OCT) and optical coherence microscopy (OCM) are used for *in vivo* visualization of plant tissue (Hittinger et al 2000, Sapozhmikan et al., 2004). In the last decade, OCT had been extensively used in medical studies to see the epithelium, connective tissue and their internal structures (glands, vessels) (Sergeev et al., 1997, Gladkova et al., 1999, Kamensky et al., 1999). Many pathological processes (such as dysplasia and cancer), alter typical structures of tissue and these alterations can be detected by OCT (Gladkova et al., 1999, Sergeev et al., 1997). This imaging technique ensures high resolution and performs *in vivo* observation (monitoring) of the physiological state and features of intra-tissular structures in humans (Kamensky et al., 1999). Lately, the OCT technique has been used to study

Institute of Applied Physics, Russian Academy of Sciences, Nizhny Novgorod, 603950, Russia, University of Texas of Medical Branch (UTMB), Galveston, 77550, USA. E-mail: vsapozh@utmb.edu, vvsap1972@mail.ru plant tissues and monitoring of intratissular changes in plants (Sapozhnikova et al., 2004). The non-destructive image tissue control performed in real time might be helpful to evaluate the ecophysiological state of plants affected by external and internal factors. This information is necessary for deeper understanding of the mechanisms of plant adaptation under influence of changing environmental conditions.

Sufficient water supply is necessary for normal functioning of plants. The problem of water supply regulation is especially crucial in urban areas, where, plants often suffer from high salt concentrations and industrial pollution. These environmental factors are leading to qualitative and quantitative modifications in the energy and structural cell metabolism. Metabolic modifications leads to insufficient production of energy, which results in damage of plant water exchange.

In this chapter, we report the first results on the use of OCT imaging for water absorption monitoring in plants. The process is very significant in allelopathic relations. The aim of this study was to evaluate the feasibility of OCT to visualize internal structure of plants and monitor intra-tissular alterations caused by external stress (water deficit). We also studied the differences in topography of water absorption between control seeds and seeds exposed to gradient of magnetic field during the first hour after turgescence started.

2. OPTICAL COHERENOE TOMOGRAPHY (OCT)

2.1 Visualization of Plant Tissue

Principle: For a long time, optical methods were not used to see the structural features of biological tissues due to their low transparence in the visible and near-IR wavelength ranges. However in range 0.8-1.3 μm biological tissues are optically turbid mediums, they do not greatly absorb light, but highly backscatter simultaneously. The idea of imaging in optically turbid media relies on temporal selection of the scattered component of a probing light. In OCT the selection is performed by using optical interferometry with broadband light sources in the near-IR range. A fraction of light penetrates into certain depths of the medium, reflects from tissue optical imperfections and retains coherent properties with the probing beam. In the reference arm, light reflects from a mirror to interfere with the backscattered light from the object under investigation. Interferometric fringes occur only if the optical distances in arms of interferometer match to within the coherence length of the light source. Therefore, axial (in-depth) OCT resolution is determined by the coherence length of the light source and is usually 10-15 μm (Gelikonov et al., 2003; Schmitt 1999).

Materials required: OCT setup, *Tradescantia pallida* (Rose), microscope, object-plates, cover glasses.

Procedure: OCT studies were carried out using a portable 40x40x15 cm fiber-optic tomograph (IAP RAS, Nizhny Novgorod, Russia). A schematic of the OCT device is shown in Fig. 1. Light from a superluminescent diode centered at 1,300 nm and having a bandwidth of 42 nm was combined in a multiplexer in a single optical path with a red laser used for alignment purposes. A 50/50 fiber coupler was used to split light between the sample and reference arms of the fiber-optic interferometer. Light returned from the sample arm of the interferometer was combined with light returned from the reference arm and generated an interferometric signal only when the distance to scattering in the sample matched the reference arm length to within the source coherence length. The value of the coherence length that determines the longitudinal (in depth) resolution of the OCT setup was measured to be 15 μm. The transverse resolution determined by beam waste on the plant was selected to be the same as the longitudinal resolution. We measured backscatter from the sample by changing the optical path with the aid of a longitudinal piezo-scanner and by detecting the envelope of the interferometric signal at the Doppler frequency (700 kHz). Then the signal was transmitted to a standard PC through an analogue-to-digital converter. Cross-sectional images on the display were built-up from multiple axial scans at different transverse positions in the sample. The transverse position of the scanner was monitored by the computer through the digital-to-analog converter. The transverse range of scanning was about 2 mm. The acquisition time for one tomogram with 200x200 pixels was about 1 s. A flexible miniature probe (2.8 mm outer dia) allowed data to be acquired from any part of a plant *in situ*. A series of experiments were conducted on this setup with the aim to study the internal structure of a plant. A miniature probe was used, permitting easy access to different parts of the plant. The OCT images were verified with standard microscopy. Microscopy sections were taken from the same regions where OCT images have been obtained. The object of investigation was *Tradescantia* grown at 20°C in peat soil at 12 h light and intensive daily watering. We chose this plant because its structural features of the leaf surface are very convenient for detection with OCT. *Tradescantia* has large watered superficial cells, therefore, all structural alterations caused by changes in water content of leaves are sure to be clearly detected with OCT.

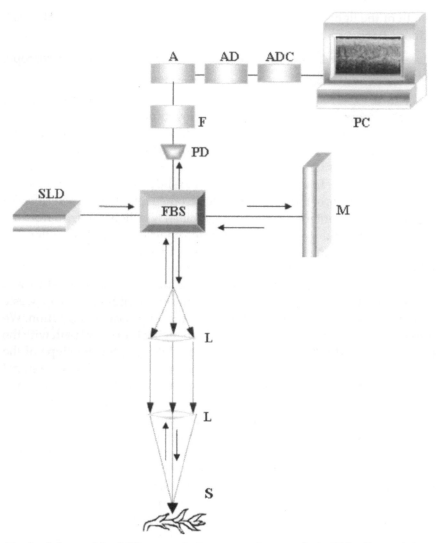

Fig. 1 Scheme of the OCT system. SLD—superluminescent diode; FBS—fiber-optic beam splitter; M—mirror; L—lenses; S—sample; PD—photo-diode; F—band-pass filter; A—logarithmic amplifier; AD—amplitude detector; ADC—analog-to-digital converter; PC—computer.

Results: In the first series of our experiments, we studied the capabilities of optical methods to visualize internal structures of plants. OCT images of the plant show the capability of OCT to identify tissue structures at depths of 1.5-2 mm. Individual cells are clearly distinguished due to the difference in scattering properties of their structural elements. Cellular walls, for instance with a layered fibrillar structure, can scatter incident light in a greater degree than the water-filled cell contents. Figure 2a and 2b shows *in vivo* images of a region of *Tradescantia* leaf acquired by OCT and microscopy with the same magnification. In the tomograms, the hexagonal, densely adjoining one another transparent cells of the leaf epidermis are clearly distinguished. The epidermal layer can be clearly detected by OCT due to heavy watering of cells of this tissue type. On the microscopy section and in the OCT image, one can see that the epidermis consists of one layer of colourless water-filled cells. Analysis of OCT and histological data shows that under the layer of epidermal cells, there is a layer of photosynthetic parenchyma. In some images, it is possible to trace regions of vascular bundles, which are not clearly detected by OCT (Fig. 2a, 2c).

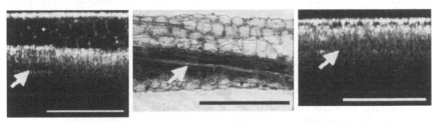

Fig. 2   OCT image and microscopy section of a plant. **a** – OCT image of the leaf from upper side; **b** – microscopic section of the leaf; **c** – OCT image of lower side of the leaf. White bar in the images correspond to 1 mm. Arrows indicate vascular bundles in OCT images and in microscopy sections

Limitations   OCT method has relatively insufficient spatial resolution in comparison with light microscopy.

2.2   Monitoring of Plant Water Content

Principle: Non-invasive visual inspection of plant tissue *in vivo* may be useful to evaluate the ecophysiological state of plants under the influence of external and internal factors. This information is indispensable for a deeper understanding of the mechanisms of plant adaptation to changing ecological conditions. One of the key requirements for normal functioning of plants is sufficient water supply. Salinity is a major issue for irrigated agricultural lands as well as for coastal areas. The problem of water supply regulation is also very important in urban areas, where, plants often suffer from above normal salinity and industrial pollution leading to qualitative and quantitative changes in energy and constitutive exchange. These modifications manifest themselves as insufficient energy production resulting in the failure of water exchange in plants. In this chapter we report the first results on the use of OCT imaging for monitoring the physiological and morphological states of plant tissues under the influence of external factors. This study is aimed at evaluating the feasibility of OCT imaging for visualization of plants and monitoring of the structural and functional states of plant tissues under external stresses such as water shortage.

Materials required: OCT setup, *Tradescantia pallida* (Rose) weighing machine

Procedure: Drought was modelled by no watering during three days before the experiment. Roots were previously removed from the soil and the plant was placed in a dry warm room at a temperature of 25-27°Ñ and humidity of 40-50%. After simulated water deficiency, plants were placed in water. OCT images of leaf tissues of one and the same plant were obtained under normal conditions and after three days simulated drought. Then OCT images were taken every five minutes during two hours after the plant had been placed in water. Final OCT images were obtained next day.

Results: In the second series of our investigation, we studied the feasibility of using OCT for *in vivo* monitoring for the dynamics of plant tissue water adsorption, under influence of water deficiency. The results of the OCT monitoring of water absorption of leaf tissue are illustrated in Fig. 3. Knowing changes in the size of cells, we can determine the total loss of water in the plant leaf. When comparing leaf surface of a plant grown at normal watering and of the same plant after water deficit, one can see a considerable difference in the thickness of the epidermal and palisade layers. The thickness of both the entire leaf and the epidermal and palisade layers is changed by approximately 3.5 times. The analysis of sizes of cells and of the whole epidermal layer of the leaf before and after simulated drought has shown that the plant lost up to 60% of water. These results correlate with data obtained with weight method (up to 50% weight lost in the process of tissue dehydration). It is noticeable that during wilting the cell size shrinks by 2.5 times on an average. More significantly water is lost by cells of the outer upper epidermal layer. Normally 2-3 cellular layers of the palisade are seen in OCT images, but after drought OCT detects only one or sometimes two layers.

OCT images of plant tissue in the process of its saturation with water are shown in Fig. 3b-3i. It can be seen from this figure that the thickness of the epidermal and palisade layers increase linearly with the time of the plant being in water. The layers reach their maximum thickness as soon as 40 minutes after the plant has been placed in water (Fig. 3g). Correspondingly, during the same time an increase in the volume of cells of the upper epidermal layer situated closer to the surface is observed. Most significantly the cells and the epidermis increase within the first 5-25 minutes (Fig. 3b-3f)

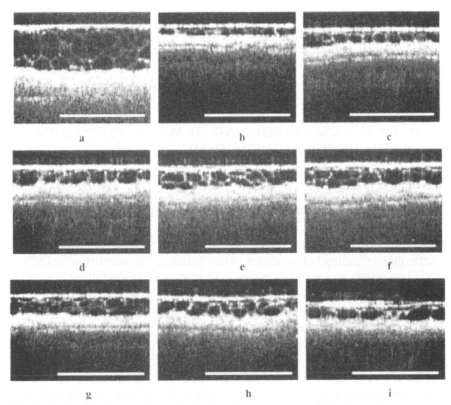

Fig. 3 Monitoring of water absorption in plant tissue: **a** – norm, **b** –3 days without water, **c** – 5 minutes in water, **d** - 10 minutes in water, **e** – 15 minutes in water, **f** - 25 minutes in water, **g** - 40 minutes in water, **h** – 2 hours in water, **i** –in water on the next day.

in the process of water saturation. Then the rate of water saturation considerably decreases. Within the following 80 minutes, no changes in the epidermal thickness are observed. Apparently, water saturation reaches its limit at this stage of plant regeneration after drought (Fig. 3g-3h). OCT images of *Tradescantia* leaf surface acquired 27 hour after the plant has been placed into water, do not differ from those of the plant leaf placed in water for 2 hour. The thickness of the epidermal layer and of the cells themselves almost did not change either in two hour or in 27 hour after watering (Fig. 3i). OCT can clearly visualize water saturation of upper epidermal cells. In Fig. 3, one can see that as the plant absorbs and saturates water, outer cellular layers of the leaf epidermis appear. The acquired OCT images distinctly show how significantly linear dimensions of the epidermis and palisade can change due to water absorption or desorption by the plant. When water amount has decreased, the size of the epidermal and palisade cells are changed. First the outer cellular layers closer to the surface are dehydrated. After drought, water absorption is accompanied by the appearance of outer cellular layers of the epidermis. This process is well defined by OCT within the first 30 minutes. Further OCT monitoring revealed no changes in the size of the epidermis and palisade and their individual cells. Next day, the hydrostatic pressure (turgor) of the plant was not totally restored.

3. OCT AND OCM MONITORING OF WATER ABSORPTION BY SEED UNDER MAGNETIC FIELD

Principle: Water regime is closely related to the whole complex of vital functions of seeds. One of the main factors governing seed germination is water absorption and associated turgescence of parts of the seed, as well as structural and metabolic changes. The mechanism of water penetration into seeds appeared to be most controversial because there are certain technical difficulties of indirect observation of this process. The non-invasive OCT or OCM imaging can provide a *in vivo* map of structural alterations occurring in subsurface layers of seeds during water absorption.

Pre-sowing stimulation of seeds by low-intensity physical factors aimed at improving their sowing qualities (germination energy, field germination power and eventually yield and their quality) have been successfully used in agriculture. One of the effective and practically proven methods is treatment of seeds by low-intensity gradient of magnetic field (GMF). But standard procedure for revealing an optimal treatment mode including estimation of germination energy and germination power can takes up to 7 days. Therefore, fast revelation an optimal mode of GMF treatment is still an actual problem.

Material required: OCT and OCM setups, pulsed GMF with pulse duration 0.1 s and maximum amplitude 5 mT, wheat seeds (*Triticum L.*) 'Moskovskaya-35', barley seeds (*Hordeum L.*) 'Zaozersky-85', long-fibred flax seeds (*Linum usitatissimum L.*) and cucumber seeds (*Cucumis sativus L.*) 'Kustovoy'.

Procedure: The next series of experiments were performed on seeds. The investigation aimed at imaging of structural alterations occurring in subsurface layers of seeds during the first hour of watering. The investigation of seeds turgescence was performed on wheat seeds (*Triticum L.*) 'Moskovskaya-35', barley seeds (*Hordeum L.*) 'Zaozersky-85', long-fibred flax seeds (*Linum usitatissimum L.*) and cucumber seeds (*Cucumis sativus L.*) 'Kustovoy'. Groups of barley seeds, cucumber seeds and long-fibred flax seeds were exposed to a 'mild' (low-intensity) physical factor (the pulsed GMF with pulse duration 0.1 s and maximum amplitude 5 mT). The 4 successive pulses were administrated during 0.4 s. The control group included seeds which were not affected by GMF. Control seeds and GMF seeds were soaked simultaneously after 41 hours of GMF treatment. OCT images were obtained each 5 minutes during the first hour after the process of turgescence of seeds started. For this experiment OCM set up with the following parameters was used: spatial resolution was 3-4 μm, depth of penetration was 1-2 mm, image acquisition time was 1-3 s (1).

Results: Typical OCT images of dry barley seeds out of the control and GMF groups are shown in Figures 4a and 4b. A highly scattering layer with thickness of about 100 μm is clearly seen on the both images. No differences between the control and GMF seeds are observed.

Experiment 1. *Dry Barley seeds*
Figures 4c and 4d demonstrate OCT images of two seeds out of the GMF group after 60 minutes when turgescence has started. One can distinctly detect darker layers of watered zones and the heterogeneous zones of water absorption. The water absorption zones merged in a united system of microcapillary vessels with sizes 50-10 μm. Individual differences in the structure of the water absorption zones in the seeds are also clearly seen.

Figures 4e and 4f show OCT images of two control seeds after 60 minutes when turgescence has started. Similar to the GMF seeds, individual structural differences of the seeds are clearly visible here. However, after the same time period the heterogeneous absorption zones (Fig. 4f) are less expressed than in the GMF seeds (Fig. 4d). The bright area corresponding to highly scattering regions (Fig. 4d) is narrower (about 100 im) in the control than in GMF seeds (about 200 μm). Thus OCT imaging of barley seeds can distinctly visualize water absorption processes within the first hour, as well as, individual variations in different seeds. The variations reflect the phenomenon of biological variability of seeds at the tissue level.

Experiment 2. *Flax seeds*
OCT studies demonstrate that seeds of other type also have individual morphological differences, which are reliably detectable by OCT even in dry seeds. Figures 5a and 5b display OCT images of two different long-fibred flax seeds (control group). In Fig. 5a, a highly scattering area with a thickness of 80-100 μm is clearly distinguished, whereas, in Fig. 5b there is one more layer at a depth 200 μm, which is weaker expressed in Fig. 5a.

Five minutes later, a well-structured water distribution becomes apparent in the OCT image of the flax seed. The upper bright layer is separated by a darker layer from a highly scattering area (about 50 μm). The darker layer has a thickness of (about 40 μm). Below there is a water absorption area

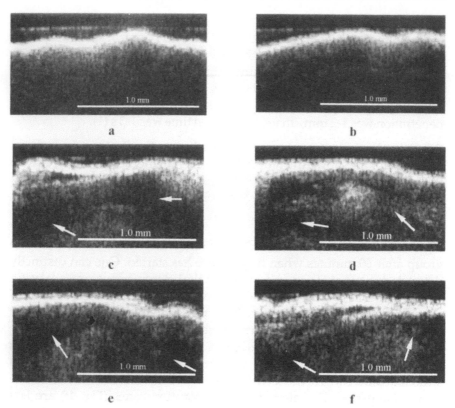

Fig. 4 a - dry barley seed (control group); **b** - dry barley seed (GMF group); **c, d** – 2 different seeds from GMF group after 60 minutes of turgescence. Arrows indicate microcapillary vessels; **e, f** – 2 different seeds from the control group (the same time period).

(100-150 μm) with clearly detectable water transport channels about 40 μm in diameter, which are occasionally merged in groups (Fig. 5d, right).

Figures 5e and 5f show OCT images of two flax seeds (GMF) 5 minutes after being soaked in water. Multiple grouping of water transport channels with formation of lens-shaped water-bearing structures can be clearly distinguished.

Experiment 3. *Cucumber seeds*
Figure 6 demonstrates OCT images of cucumber seeds, showing most pronounced differences in topography of water absorbing structures among all seeds being studied. An OCT images of dry cucumber seeds of control and GMF groups are presented in Fig. 6a and 6b. No visual differences between control and GMF seeds are observed. In this figure a highly

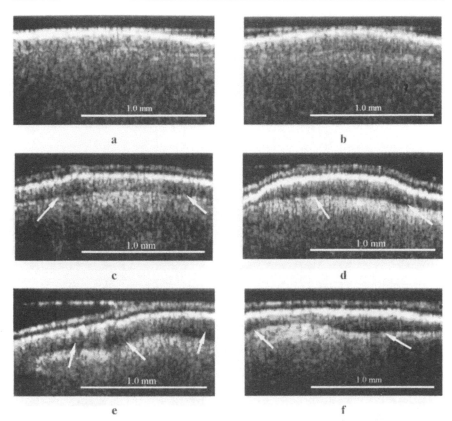

Fig. 5 a, b – 2 different seeds of long-fibred flax (control group); **c, d** – 2 seeds of the control group after 5 minutes in water; **e, f** – 2 seeds of GMF group after 5 minutes in water. Arrows indicate lens-shaped water-bearing structures. Water transport channels are below them.

scattering layer with a total thickness of about 100 μm is clearly seen. The layer is separated into two sub-layers of equal thickness.

After 5 minutes: However after 5 minutes soaking, a distinct structuring of the upper highly scattering layer in GMF seeds is observed (Fig. 6c). The formation of lens-shaped water absorbing structures analogous to those in flax seeds is also detected. During the same time period (5 minutes after soaking – Fig. 6d) in control seeds the differentiation of the highly scattering layer into two sub-layers is less pronounced and there are no lens-shaped water absorbing structures.

After 10 minutes: Ten minutes after the water absorption has started, the process of differentiation of water absorbing layers is faster in GMF seeds (Fig. 6e) than in control seeds (Fig. 6f). The highly scattering layers are larger,

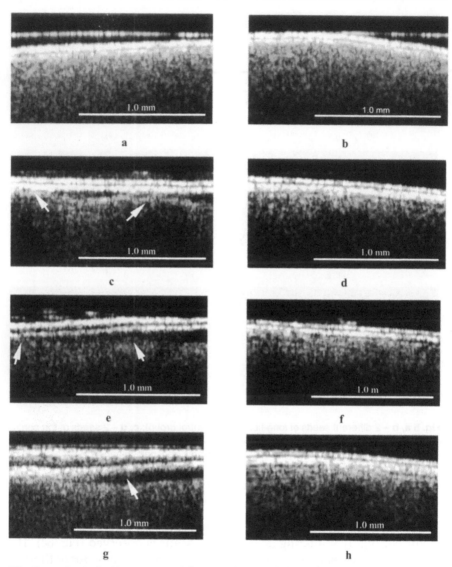

Fig. 6.   **a, b** - typical tomograms of dry cucumber seed of control and GMF groups; **c, e, g** – cucumber seeds of GMF group after 5, 10 and 15 minutes in water accordingly; **d, f, h** – cucumber seeds of the control group (the same time periods). Arrows indicate lens-shaped water absorbing structures.

the separating dark layer is thicker and the degree of formation of the lens-shaped water absorbing structure is greater. In control seeds (Fig. 6f), after 5 minutes of the beginning of water absorption the differentiation of water absorbing structures has changed insignificantly (Fig. 6d).

After 15 minutes: After 15 minutes of soaking GMF seeds in water (Fig. 6g), the differentiation of water absorbing layers is still greater than in control seeds (Fig. 6h). Thus OCT can clearly visualize both individual differences in morphology of water absorbing structures in seeds, as well as changes in the dynamics of differentiating of these structures.

Experiment 4. *Wheat seed.*

OCM provides information on processes of seed watering at the cellular-tissue level with higher spatial resolution than OCT. This can be illustrated by an OCM image of a dry wheat seed (Fig. 7a). A thin layer of pericarp and seedcoat with a total thickness of 40 μm and an underlying aleuron layer with a thickness of 60 μm are clearly visible. The dark layer in the dry wheat seeds indicates that the probing light does not penetrate into a depth more than 100 μm. After 12 hours of water absorption, the penetration depth of probing light is about 250 μm (Fig. 7b). At this time, the pericarp and seedcoat appear in the images as two clearly delineated layers with a total thickness of about 60 μm. The aleuron layer with large cubical cells and underlying layers with alternating dark and light regions are clearly seen in the image.

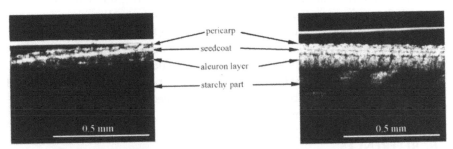

Fig. 7 OCM images of wheat seed **a** – dry; **b** – after 12 hours in water.

4. DISADVANTAGES

(i). OCT method has worse spatial resolution in comparison to destructive light microscopy.

(ii) The imaging depth of 1-1.5 mm in OCT and OCM techniques is insufficient to visualize tissues through the whole plant.

5. CONCLUSIONS

To investigate physiological reactions occurring in a plant, as it grows and develops or under the action of environmental external and internal factors, researchers employ methods that subjects the plants being studied to stresses. This inevitably has an influence on the final results of such experiments. OCT imaging can quickly within 1-3seconds, detect physiological activity of plants in the process of their growth and development with no external influence on the plant. In this chapter, we have reported the results of visualization of plant tissue and monitoring of water saturation of dried plants.

OCT images were compared with standard microscopy sections. We showed that OCT imaging can be used for *in vivo* evaluation of plant cells sizes and for monitoring of plant morphological and physiological states under different external factors. Since individual structures of biotissue have different scattering parameters, the borders between them are clearly seen in OCT images. The OCT can also be used to study the morphological and physiological states of plants that grow under unfavorable ecological conditions and under the influence of external chemical and physical factors.

Optical Coherence Tomography and Optical Coherence Microscopy can be used for *in vivo* control of turgescence of native seeds without staining and contrast agents. We demonstrate that seeds have individual morphological differences, which are reliably detected by OCT even in dry seeds. Thus we can clearly visualize both individual differences in morphology of water absorbing structures in seeds, as well as changes in the dynamics of differentiating of these structures. OCT enables monitoring of water absorption in seeds. It was found that the water absorption rate differs in control seeds and in seeds exposed to weak gradient magnetic field. Therefore, we believe that OCT in the future can be used for fast revealing of optimal pre-sowing treatment of agricultural seeds under influence of low-intensity physical factors.

6. REFERENCES

Gelikonov,V.G., Gelikonov, G.V., Ksenofontov, S.Yu., Kuranov, R.V., Morozov, A.N., Myakov, A.V, Turkin, A.A., Turchin, I.V. and Shabanov, D.V. (2003). New approaches in broadband fiber-optical interferometry for optical coherent tomography *Radiophysics and Quantum Electronics* **46** : 550-564.

Gladkova, N.D., Shakhova, N.M., Shakhov, A.V., Petrova, G.A., Zagainova, E.V., Snopova, L.B., Kuznetzova, I.A., Chumakov, Y.P., Feldchtein, F.I., Gelikonov, V.M., Gelikonov, G.V., Kamensky, V.A., Kuranov, R.V. and Sergeev, A.M. (1999). OCT monitoring of the pathophysiological processes. In *Proceedings. SPIE* **3598**, pp. 86-97. San Jose, USA.

Haseloff, J. (1999). GFP variants for multispectral imaging of living cells. *Methods in Cell Biology* **58**: 139-151.

Hittinger, J.W., Mattozzi, M., Myers, W.R., Williams, M.E., Reeves, A., Parsons, R.L., Haskell, R.C., Petersen, D.C., Wang, R. and Medford, J.I. (2000). Optical

Coherence Microscopy. A technology for rapid, *in vivo*, non-destructive visualization of plant and plant cells. *Plant Physiology* **123**: 3-15.

Kamensky, V.A., Feldchtein, F.I., Gelikonov, V.M., Snopova, L.B., Muraviov, S.V., Malyshev, A.Y., Bityurin, N.M. and Sergeev, A.M. (1999). *In situ* monitoring of laser modification process in human cataractous lens and porcine cornea using coherence tomography. *Journal of Biomedical Optics* **4**: 137-143.

Sapozhnikova, V.V., Kamensky, V.A., Kuranov, R.V., Kutis, I., Snopova, L.B. and Myakov, A.V. (2004). *In vivo* visualisation of *Tradescantia* leaf tissue and monitoring the physiological and morphological states under different water supply conditions using optical coherence tomography. *Planta* **219**: 601-609.

Schmitt, J.M. (1999). Optical Coherence Tomography (OCT): A Review. *IEEE Journal on Select Topics In Quantum Electronics* **5**: 1205-1215.

Sergeev, A.M., Gelikonov, V.M., Gelikonov, G.V., Felchtein, F.I., Gladkova, N.D., Shahova, N.M., Snopova, L.B., Shakhov, A.V., Kusnetzova, I.A., Denisenko, A.N., Pochinko, V.V., Chumakov, Y.P. and Strelzova, O.S. (1997). *In Vivo* endoscopic OCT imaging of precancer and cancer states of human mucosa. *Optic Express* **1** (**13**): 432-440.

Chapter

Optical Coherence Microscopy: Study of Plant Secretory Structures

V.V.*, Roshchina, I.S. Kutis**, L.M. Kutis***,
and G.V. Gelikonov** and V.A. Kamensky**

1. INTRODUCTION

Secretory cells filled with allelochemicals participate in the allelopathic relations. Their structure may be changed under various external factors that needs special methods for the observations. The application of non-invasive optical techniques to diagnose the physiological state of plants is one of the major problems of plant ecophysiology.

Appearance of broadband (femtocorrelated) sources of radiation such as superradiant diodes and femtosecond lasers permits us to see the images of biological objects' architectonics using optical techniques. Among the methods, optical coherence tomography (OCT) based on a biotissue probing by femtocorrelated infrared radiation and interference recording of near-infrared light backscattered from the tissue is recommended that allows a visualization of internal structure in biological objects with high resolution in the real time. Optical coherence microscope (OCM) is a new device that combines merits of confocal microscope and optical coherence tomograph, the two well-known modalities used to examine the objects in strongly scattering light hence, optically nontransparent media, in particular biological tissues (see schemes in chapter 6). This microscope provides both,

*Institute of Cell Biophysics, Pushchino,
** Institute of Applied Physics, Russian Academy of Sciences, 46 Ulyanov Str., 603950 Nizhny Novgorod tel. +7(8312)368010, E-mail: vlad@ufp.appl.sci-nnov.ru,
*** University of Nizhny Novgorod, Department of Biology, 23 Gagarin Ave., 603091 Nizhny Novgorod, Russia

geometric (confocal) and temporal (coherent) selection of information signal. These vital methods without any fixation are successfully employed in medicine (Boppart et al., 1999) and already used for investigation of different types of plant structures (Sapozhnikova et al., 2004). This technique could be applied to the study of secretory cells of allelopathically active plants.

2. TECHNIQUE OF OPTICAL COHERENT MICROSCOPY

Principle: Optical Coherent Microscopy, using femtocorrelated infrared radiation and interference recording of near-infrared light backscattered from the tissue, allows us to see the internal structure in biological objects with high resolution.

Materials: Optical coherence microscope (OCM), personal computer for visualization and recording of images, distilled water, slides and cultivated plants in natural conditions or freshly harvested. The following plants were investigated: *Nicotiana affinis* L., *Calendula officinalis* L., *Urtica dioica* L., *Atriplex* sp., *Picea* and *Pinus sylvestris*.

Apparatus: A portable OCM [developed and fabricated at the Institute of Applied Physics, Russian Academy of Sciences, Nizhny Novgorod] was used. A broadband interferometer was used as a receiving OCM system and combined radiation of two superradiant semiconductor diodes with central wavelengths of 907 nm and 948 nm as a source. The radiating power at the object was about 1 mW; spatial resolution was less than 5 mcm. The two-dimensional field obtained at scanning in depth (by changing optical path difference in the interferometer arms) and in width (by moving the probing radiation beam over the surface of the object) was visualized on the PC monitor. The time of acquisition of a single 400x400 pixel image was 3 to 4.5 sec. The probing depth did not exceed 400-500 mcm on the average, 256 colours varying from black to white with intermediate grades of grey were used in the OCM images. The darker sections of images correspond to higher signal intensity and the lighter ones to lower intensity. Consequently, structures with stronger scattering of probing radiation are dark and weakly scattering structures are light.

Procedure: Water was taken as an immersion medium. The object under study was placed on a wet specimen stage so that the section of its surface to be investigated should be above the scanning zone and was clamped by a slide. After that OCM images of the biological object were acquired. The images were recorded and stored in the personal computer for further analysis.

2.1 Images of Secretory Hairs

Experiment 1. *The secretory trichomes of allelopathically active species.*
OCM images of trichomes on the leaves of *Calendula officinalis* L. and *Nicotiana affinis* L. or trichomes and stinging hairs of *Urtica dioica* L. are shown in Figs 1 and 2. Dark sports and lines are the position of dense material, which is contained in lignin-impregnated cell walls and in various crystal inclusions. Crystals are seen as dark structures, especially in stinging hairs. Such crystals as secretions, mainly of calcium or silica salts, may include phenols and alkaloids or other compounds as secretions (Vasiliev, 1977; Fahn, 1979; Roshchina and Roshchina, 1993). Light (white) spaces of the hair were filled with liquids. Only small crystals of sesquiterpene lactones (in *Calendula*) or alkaloids (in *Nicotiana*) may be deposited on the cell surface. As for stinging hair of *Urtica dioica*, stinging silica crystal is seen on the tip of the trichome and after a contact with animal skin, it falls off. Besides stinging hair on the surface of *Urtica* are observed as ordinary small hair, whose structure differs from stinging long trichomes. Within stinging hair, there are numerous compounds viz., acetylcholine, serotonin, histamine, formic acid and others, which act as allelochemicals (Roshchina and Roshchina, 1993).

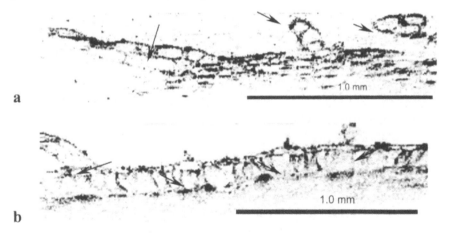

Fig. 1 Leaf trichomes: **a** – *Calendula officinalis* L., **b** – *Nicotiana affinis* L.

2.2 Images of Resin Ducts

The resins of conifer plants contain terpenoids and phenols, which demonstrate allelopathic effects.

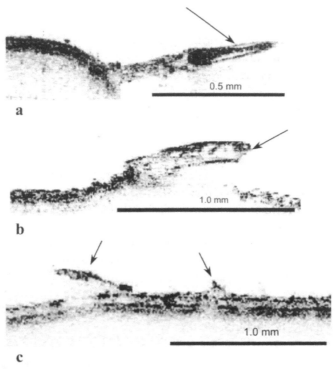

Fig. 2 OCM images of trichomes (shown by arrows) on the *Urtica dioica* L. leaf: **a** – sting with silicon tip (edge); **b** – without tip (edge); **c** – ordinary trichome.

Experiment 2. OCM *images of resin ducts.*

OCM images of resin ducts in in *Picea* and *Pinus silvestris* L. are shown in Fig. 3. The ducts appear as light spots located near the needle surface. Resin if not crystallized, is in a liquid form and is not registered as a dense material.

2.3   Images of Salt Glands

Salts are present in secretions of many plants, especially growing in arid regions (Roshchina and Roshchina, 1993). OCM images of similar salt secretory structures were received (Fig.4) as per Experiment 3.

Experiment 3. OCM *images of salt secretory strucrures.*

Salt glands of plants from *Atriplex* genus contain inclusions in the form of crystals of siliceous or sulphate salts of calcium and magnesium (Fahn, 1979). Usually the crystal particles also include phenols (see Chapter 7). The crystals are seen as dark dense spots within the structures on OCM images of the optical slices from the gland (Fig. 4). Profiles of signal intensity along

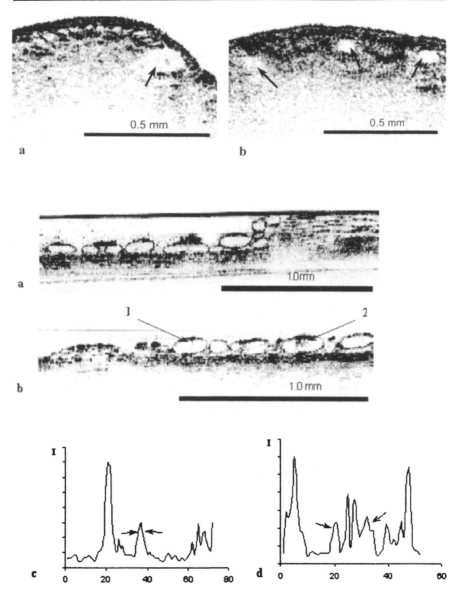

Figure 4. a, b – Two different sections of the lower part of the *Atriplex sp* leaf; **c, d** – profiles of tomographic signal intensity for two inclusions (Fig. 4b, shown by arrows). The tomographic signal intensity profiles are given along the section across gland walls and inclusions. The distance (in mcm) from the tissue surface is laid off along the X-coordinate, and the intensity along the Y-coordinate. The peaks corresponding to salt gland inclusions are shown by arrows. Profile **c** corresponds to inclusion 1, and profile **d** to inclusion 2 in Fig. **4b**.

sections across both the gland walls and across the inclusions are presented. The plants with salt glands demonstrate great tolerance to water deficit in arid regions and they are pioneers in waste lands. Allelochemicals of the plants (e.g., phenols) may be included into the salt crystals. In this concentrated form, they act more effectively against neighbours.

Adavantages
  (i) Optical coherence microscopy (OCM) with resolution not less than 5 mm allows observation of structures in biological tissues *in vivo* at the cellular level.
 (ii) No fixators or fine sections of the studied object are required.
(iii) Changes in the dense secretory structures occurred due to allelochemical effects may be analyzed with the technique.

3. REFERENCES

Boppart, S.A., Goodman, A., Libus, J.J., Pitris, C., Jesser, C.A., Brezinski, M.E. and Fujimoto, J.G. (1999). High resolution imaging of endometriosis and ovarian carcinoma with optical coherence tomography: feasibility for laparoscopic-based imaging. *British Journal of Obstetrics and Gynaecology* **106**: 1071-1077.

Bouma, B.E. and Tearney, G.J. (Eds) (2002). *Handbook of Optical Coherence Tomography* Marcel Dekker: New York., USA.

Fahn, A. (1979). *Secretory Tissues in Plants.* Academic Press, London,UK.

Gelikonov, V.G., Gelikonov, G.V., Ksenofontov, S.Yu., Kuranov, R.V., Morozov, A.N., Myakov, A.V, Turkin, A.A., Turchin, I.V. and Shabanov, D.V. (2003). New approaches in broadband fiber-optical interferometry for optical coherent tomography. *Radiophysics and Quantum Electronics* **46**: 550-564.

Hittinger, J.W., Mattozzi, M., Myers, W.R., Williams, M.E., Reeves, A., Parsons, R.L., Haskell, R.C., Petersen, D.C., Wang, R. and Medford, J.I. (2000). Optical Coherence Microscopy: A technology for rapid *in vivo*, non-destructive visualization of plant and plant cells. *Plant Physiology* **123**: 3-15.

Roshchina, V.V. and Roshchina, V.D. (1993). *The Excretory Functions of Higher Plants.* Springer-Verlag, Berlin., Germany.

Sapozhnikova, V.V., Kamensky, V.A., Kuranov, R.V., Kutis, I.S., Snopova, L.B. and Mjakov, A.V. (2004). *In vivo* visualization of *Tradescantia* leaf tissue and monitoring the physiological and morphological states under different water supply conditions using optical coherence tomography. *Planta* **219**: 6001-609.

Schmitt, J.M. (1999). Optical Coherence Tomography (OCT): A Review. *Institute of Electric and Electronic Engineers Journal of Selected Topics in Quantitative Electronics* **5**: 1205-1215.

Shakhova, N.M., Gelikonov, V.M., Kamensky, V.A. and Kuranov, R.V. (2002). Clinical aspects of the endoscopic optical coherence tomography and the ways for improving its diagnostic value. *Laser Physics* **12**: 617–626.

Vasiliev, A.V. (1977). *Functional Morphology of Secretory Plant Cells.* Nauka, Leningrad, Russia.

Chapter

Laser-scanning Confocal Microscopy (LSCM): Study of Plant Secretory Cell

V.V. Roshchina, V.A. Yashin,
A.V. Kononov and A.V. Yashina

1. INTRODUCTION

Laser scanning confocal microscopy (LSCM) may produce images of high quality from fluorescing cells. The images of plant secreting cells, which excrete allelochemicals, or plant cells, which serve as acceptors of allelochemicals, may be changed in allelopathic relations that are registered by this technique. The advantages of the technique are: (i) 3-channel simultaneous detection that permits to see images excited by 3 different wavelengths light from laser and to receive common complicated interference image of the object; (ii) the possibility to have increased depth of penetration for receiving 20 visual slices (optical sections) or the complete volume (the information must be also quantitatively extracted); (iii) interchangeable filters; (iv) Graphical User Interface and production of accurate computer models as well as mathematical analysis; (v) pattern analysis of the structure.

In this chapter the application of LSCM technique to study plant secretory cells, which participate in allelopathic relations, is described.

2. LSCM TECHNIQUE

In luminescent microscope, light from the ultraviolet source (UV source) excites the fluorescence of the object, the luminescent image might be catched

Institute of Cell Biophysics, Russian Academy of Sciences, Institutskaya Str., 3, Pushchino, Moscow Region, 142290, Russia, E-mail: roshchina@icb.psn.ru directly with an eye of observer (Fig. 1). Unlike usual luminescent microscopy, confocal microscope has a special confocal aperture (pinhole), from which a fluorescence of the object excited with laser beam with a certain wavelength passes and multiples by photomultiplier before the eye visualization. Construction of aperture permits to focus the light beam on the different depths of the object.

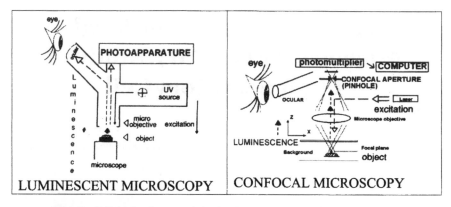

Fig. 1 Principal schemes of the luminescent and confocal microscopy

Principle: Well-seen image of a cellular structure and its optical slices show the changes induced by many experimental factors, including allelochemical testing.

Materials required: Carl Zeiss Laser Scanning Confocal Microscope LSM 510 NLO Carl 'Zeiss', object glasses, cover glasses, plant allelochemicals and pigments, azulene, quercetin and rutin, rutacridone, chlorophyll, sum of carotenoids

Procedure: The fluorescence of the cells are observed and measured on the object glasses (slides). All experiments were performed at room temperature 20-22°C. The fluorescence image of the cells was also seen by the water immersion with laser scanning confocal microscope LSM 510 NLO 'Carl Zeiss'. The excitation by three types of lasers: Argon/^{2} (l 458, 488, 514 nm), HeNe1 (543 nm) and HeNe^2 (l 633 nm) were used for of the emission. In the experiments, three photomultipliers can catch the fluorescence, separately or simultaneously by use of pseudocolour effects. The image analysis was done with computer programmes LSM 510 and Lucida Analyse 5. The excitation by lasers with wavelengths 458, 488, 543 and 633 nm, the registration of the fluorescence was at 505-630 nm, 650-750 nm and 650-750 nm, relatively. Pseudocolours were according to the excitingwavelength blue for 488; green for 543 and red for 633 nm. The composite image was seen, when the images consisting of pseudocolours were superimposed and mixed.

2.1 Images of Vegetative and Generative (Pollen) Microspores

Observation. The time of observation is about 2-4 hours for vital preparation and several days for fixed preparations. All the preparations may be analyzed by 3-channel simultaneous detection to receive common complicated interference image of the object and details of its structure (see experiment 1); by receiving 20 visual slices (optical sections) or the complete volume (the information must be also quantitatively extracted) (see experiment 2); by computer modelling the images as well as mathematical analysis (experiment 3).

Experiment 1. *Three channels images of the vegetative microspores*
Figure 2 shows the LSCM views of microspores studied, excited with three different laser wavelengths. The rigid cover is well seen at the excitation 488 nm and registration at 505-635 nm. At the excitation 543 nm and 633 nm chloroplasts (red-orange colour) and other organelles (yellow colour) are observed. The fluorescence of vegetative microspores may be related to their cover (blue fluorescence of phenols) and chloroplasts (red fluorescence of chlorophyll at 650-750 nm) as shown by Roshchina et al (2004). The shift in the emitted cellular compartments dealt with the first moistening is observed as the reaction. A diffusive distribution of the elements of cover and

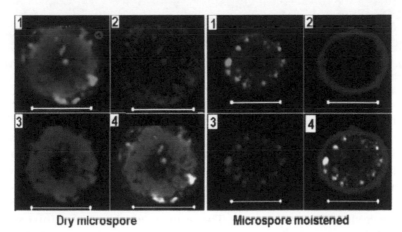

Dry microspore   Microspore moistened

Fig. 2 LSCM view of dry and moistened (15 min) fluorescent vegetative microspore of *Equisetum arvense* under three laser excitation. 1 – channel 488 nm; 2 – channel 533 nm; 3- channel 633 nm; 4-summed image with mixed (in a superposition pseudocolours. 1 bar = 20 um.

chloroplasts became very clear. The separate inclusions of the exine in the cover look brighter.

Experiment 2. *Visual slices of the vegetative microspore structure and computer modelling*

Scanning of the object along the Z-coordinate (see Fig. 1) with an interval of 1.0 µm, show slices of the microspore (Fig. 3). The slices can be collected by special computer programme for LSCM 510 and reconstruction of the separate fragment of the cell surface may be received (Fig. 4).

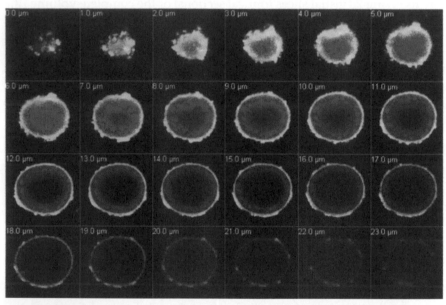

Fig. 3 Optical slices (through 1 µm) of vegetative microspore of *Equisetum arvense*. Excitement with laser beams 488 (emission > 520 nm, green pseudocolor), 543 and 633 nm (fluorescence is observed at 650-750 nm, red pseudocolor).

The first slices of microspores are the luminescent views of the cover itself (Fig. 3). Excitement with laser beams 488 nm induced the green fluorescence, which is observed at wavelengths > 520 nm. The intensive emitted inclusions are seen on the cover surface. More deep slices show red fluoresced chloroplasts (excitation by light 488, 543 and 633 nm) located on periphery in cytoplasm. After 20 µm of cutting, one can see the cover from the opposite side of the microspore. The reconstructed part of cellular surface (Fig. 4) is represented as green-fluorescent particles impregnated to red-fluorescing matrix. Pigments azulenes in a free state or a solution emit in blue, but, when they are included into cell wall, can fluoresce in red spectral

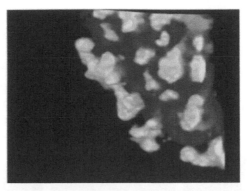

Fig. 4 Fragment of reconstructed part of cellular surface in microspore of *Equisetum arvense*

region (Roshchina, 2003). Thus, the reconstruction indicates different components of the cover.

Experiment 3. *Image of Hippeastrum hybridum pollen structure*

The analysis of LSCM images of the pollen grains of *Hippeastrum hybridum* is represented in Fig. 5. The blue and blue-green fluorescence prevail in the microspore cover, but in some internal compartments, red particles are seen (Fig. 5, left). It is known that non-mature pollen have chloroplasts, which are lost later (Roshchina, 2003). Thus, one can see the different pollen population with whole chloroplasts and disappearing plastids. The optical slices of wet, swelled pollen (Fig.5, right) indicate, mainly, blue-green emission, especially in the cover.

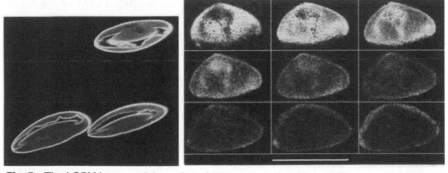

Fig. 5 The LCSM images of the pollen of *Hippeastrum hybridum*. Left – common image of dry pollen (excitation by laser beam 458 nm, emission > 520 nm and laser beam 633 nm, emission >670 nm), Right – the stack of optical slices cut through 2 mm of wet pollen (excitation by laser beam 458 nm, emission at 518 nm). Bar = 50 µm

Experiment 4. *Images of developed vegetative microspores of Equisetum arvense*
As seen in Fig. 6, developing microspores of *Equisetum arvense* show various changes, depending on the accumulation of new synthesized chlorophyll (Roshchina et al., 2002; Roshchina, 2004), which fluoresce in red.

The microspore was divided after 1-2 days moistening and one of two cells lost chloroplasts, forming rhizoid. Other cell formed multicellular protallium and then thallus in 4 -7 day. The images are well seen in LSCM.

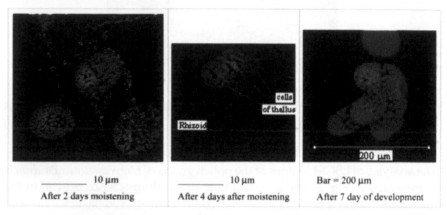

Fig. 6 Microphotograph of fluorescent developing vegetative microspore of *Equisetum arvense* under laser beam 633 nm, (emission >670 nm).

2.2   Images of Individual Substances

When light from all three channels excites the fluorescence of crystalline individual compounds such as allelochemicals flavonoids quercetin and rutin or pigments of plant cells – azulene, chlorophyll and carotenoids fluoresce in different regions of the spectra: in yellow and red or blue, red and yellow-orange, respectively (Fig. 7). It compares the light emission of the substances within cellular structures.

Experiment 5. *The fluorescence of crystalline allelochemicals and some pigments*
Yellow–green emission is peculiar to quercetin, while orange – to rutin. Azulene fluoresce in blue, carotenoids – in yellow, and chlorophyll – in red.

Experiment 6. *The LSCM image of salt gland impregnated with phenolic compounds*
Multicellular secretory structures may include various products (Roshchina and Roshchina, 1993). Some of them such as salts do not fluoresce themselves, but when included in the structures, may absorb

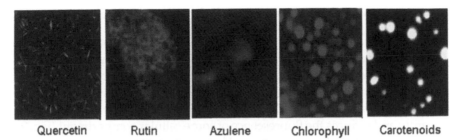

| Quercetin | Rutin | Azulene | Chlorophyll | Carotenoids |

Fig. 7 Microphotograph of the crystal compounds under three channels LSCM beams (see Procedure). Excitation by wavelengths 488, 543 nm and 633 nm. Chlorophyll or carotenoids are a sum of chlorophylls *a* and *b* or a sum of carotenoids from pea leaves purified on silicagel.

fluorescent compounds. One of the example for allelopathic active species *Chenopodium album* is shown in Fig. 8.

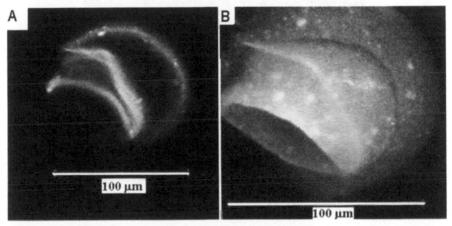

Fig. 8 The fluorescence image of *Chenopodium album* salt gland seen in laser-scanning confocal microscope. The excitation wavelength 488 nm, emission > 520 nm. A – single image of the gland slice; B – stack of the slices (sum, resulting or composite image) of the gland.

Leaf salt glands impregnated with phenolic compounds (that was identified through the staining with 2 % Ferrous chloride) can fluoresce in the structure (Fig. 8). Big crystalline bands are seen on one of the slices. Small crystals of phenols are also seen on the stack of the slices (sum image) as yellow-green lightening spots, brighter , than the other surrounded structure. Many phenolic allelochemicals may exist in such structures and affect plants grown around them, when the salt crystals fall outside leaves and fruits.

2.3 Images of Cells-acceptors Treated with Allelochemicals

Experiment 7. *Alkaloid colchicine binds tubulin in cell structures.*

Colchicine as allelochemical is peculiar to *Colchicium* genus and when it is released by the cell-donor, can interact with the cell-acceptor. This alkaloid is also known as the tubulin-binding agent, which blocks its polymerization in microtubes (Roshchina, 2005a;b). This alkaloid penetrates the cell-acceptor and induces the fluorescence of some cellular compartment (Fig. 9). The own emission of pure compound is 300 times less.

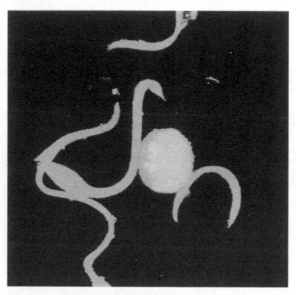

Fig. 9 The LSCM images of *Equsetum arvense* microspores stained with colchicine 10^{-7} M. The laser excitation wavelength 458 nm.

Green-yellow fluorescence of the alkaloids is seen on the cell surface and in elaters of vegetative microspores from *Equsetrum arvense* (Fig. 9) shows the presence of tubulin in the parts. The middle of the spore lights in red due to the presence of chloroplasts.

The staining of germinated pollen of *Hippeastrum hybridum* with colchicine demonstrates green-yellow emission of microtubules (better vision in black-white image) around nuclei of pollen grain (threads at the division of the nucleus) and spermium on the tip of the pollen tube, where spermium moves, as well as in some bridge sites of the tube (Fig. 10). The similar fluorescent allelochemicals may be also used as fluorescent dyes at the cellular diagnostics (Roshchina, 2005 b).

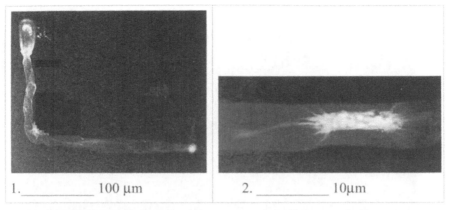

| 1. _____ 100 µm | 2. _____ 10µm |

Fig. 10 The LSCM images of *Hippeastrum hybridum* pollen tube stained with colchicine 10⁻⁷ M. The laser excitation wavelength 458 nm. 1. The bright emission is observed in nucleus of vegetative cell of pollen and in the spermium located in the tip of the tube; 2. Spermium in the tip of pollen tube. The microtubules contained of tubulin are seen.

2.4 Images of Cell-donors Releasing Secretions

Experiment 8. *The observation of the secretory hairs and their secretion*
Cell-donor of allelochemicals releases the substances out. The process may be seen with the help of LSCM technique as the study of the fluorescence of various external secretory structures. Such structures are glandular cells, which contained many potentially fluorescent substances (Roshchina and Roshchina, 1993). One of the example is shown for secretory leaf hair of allelopathically active species *Solidago virgaurea* L. (Fig.11).

Secretory hairs of allelopathic active plant *Solidago virgaurea* L. contain a lot of oils, which include terpenes. Fig. 11 (7-13) shows the fluorescent drop of the oil. Dried oil is crystallized on the surface of the secretory hair and is seen as green colour bodies. The separate components of the oil fluoresce in different spectral regions. There are three layers in the oil drops: green, yellow and red fluorescing. Perhaps, these are different components of the secretory hair. Phenols may be present in the oils of many species, besides terpeniods (Roshchina and Roshchina, 1993). Bright green lightening crystals are seen on the surface and within the hair. Oil of the medicinal species contains allelopathic active components, which can fuse into the cell-acceptor during the allelopathic relations with other plant species.

The research is supported by Grant Foundation Council, Russian Academy of Sciences "Femotosecond Optics and Physics of Superpower Laser Fields".

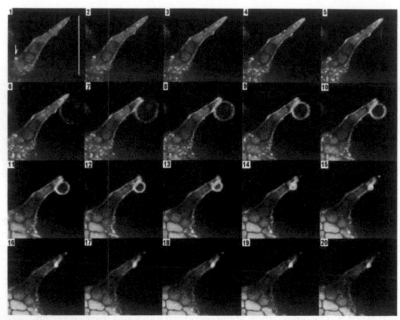

Fig. 11 The LCSM images of the secretory leaf hair of *Solidago virgaurea* L. The stack of optical slices cut through 1 mm [excitation by infrared femtosecond laser beam in track 1 800 (400) nm and in track 2 – 488 nm, emission in channel 1 400-465 nm and in channel 2 560-700 nm]. Bar on section 1 = 150 mm. Bright green lightening crystals are seen on the surface and within the hair (one of them is shown with the arrow in section 1).

3. REFERENCES

Cheng, P.C. and Summers, R.G. (1990). Image contrast in confocal light microscopy. In: *Handbook of Biological Confocal Microscopy*, J. B. Pawley (ed) pp. 179-195. Plenum Press, New York,USA.

Roshchina, V.V. (2003). Autofluorescence of plant secreting cells as a biosensor and bioindicator reaction. *Journal of Fluorescence* **13**: 403-420.

Roshchina, V.V. (2004). Cellular models to study the allelopathic mechanisms. *Allelopathy Journal* **13**: 3-16.

Roshchina V.V. (2005a). Contractile proteins in chemical signal transduction in plant microspores. *Biological Bulletin, Ser. Biol.* **3** : 281-286.

Roshchina V.V. (2005b). Allelochemicals as fluorescent markers, dyes and probes. *Allelopathy Journal* **16** : 31-46.

Roshchina, V.V. and Roshchina, V.D. (1993). *The Excretory Functions of Higher Plants*. Springer-Verlag, Berlin,Germany.

Roshchina, V.V., Yashin, V.A. and Kononov, A.V. (2004). Autofluorescence of plant microspores studied by confocal microscopy and microspectrofluorimetry. *Journal of Fluorescence* **14**: 745-750.

Salih, A., Jones, A., Bass, D. and Cox, G. (1997). Confocal imaging of exine for grass pollen analysis. *Grana* **36**: 215-224.

Chapter

Luminescent Cell Analysis in Allelopathy

V.V. Roshchina

1. INTRODUCTION

Some allelochemicals (phenols, alkaloids, terpenoids) and other biologically active compounds, can fluoresce at the excitation of ultra-violet or
violet light (Roshchina, 2002; 2003; 2004, 2005a,b). The fluorescence induced by ultra-violet or violet light has also been observed both in intact
secretory plant cells enriched in the substances and excretions from the cells.
The substance fluorescence may serve as a marker for the cytodiagnostics of
the secretory structures in luminescent microscope (they are not seen in
usual microscope without special histochemical staining). The compounds
may be also used as fluorescent probes in the studies both for the wider
laboratory practice or specially for the analysis of the mechanisms of the
allelochemical action (Roshchina, 2004). The effects of allelochemicals on
the chlorophyll fluorescence in leaves have already been described (Reigosa
Roger and Weis, 2001; Weis and Reigosa Roger, 2001)

The visible fluorescence of intact living cells is called autofluorescence. If
the cells treated with special fluorescent dyes, the cellular common
fluorescence also includes the additional light emission. Autofluorescence
could be used: (i) in express-microanalysis of the accumulation of the
secondary metabolites in secretory cells without long biochemical
procedures; (ii) in diagnostics of cellular damage and (iii) in analysis of cell–
cell interactions (Roshchina, 2003). During the plant development, secretory

Laboratory of Microspectral Analysis of Cells and Cellular Systems, Russian Academy of
Sciences, Institute of Cell Biophysics, Institutskaya str., 3, Pushchino,Moscow region,
142290, Russia, E-mail: roshchina@icb.psn.ru

cells may change both the composition of the fluorescing secretions and their amount, that could be measured by estimating their characteristic fluorescence spectra and the fluorescence intensity (Roshchina, 2003). Most medicinal plants have secretory cells, where, pharmacologically-valuable secondary metabolites are accumulated (Roshchina and Roshchina, 1993).

Fluorescence of secretory structures may be seen and measured by various technique : from simple luminescent microscope and complex laser scanning confocal microscope to microspectrofluorimetry described in this chapter.

2. VISUAL OBSERVATIONS OF SECRETORY CELLS IN LUMINESCENT MICROSCOPE

Principle: Plant enriched in secretory structures with biologically active secondary metabolites have fluorescing products in the cells and a location of the compounds could be observed using luminescent technique.

Material required: Luminescent microscope, object glasses (slices)

Procedure: The fluorescence of the objects, which lie on the object glasses (slices) and excited by ultra-violet (360-380 nm) or violet (400-420 nm) light, may be seen in luminescent microscope with multiplication of objectives x 10, 20, 40 or with water immersion x 85, or with immersion oil x 60 and 85 .

Observation: Depending on the purpose, the fresh plant samples on slices may be observed immediately or after the histochemical staining with following analysis.

Experiment 1. *Visual observation of secretory cells in luminescence microscope*
Various types of secretory cells in allelopathically active plants contain fluorescing secondary products (Fig. l). Secretions from the above ground parts of plant (in leaves, flowers, stems) were concentrated in secretory hairs and glands. Whereas secretions of roots are in secretory reservoirs and idioblasts (ordinary cells which accumulate secretory products) or may be released by the secretory surface of the root tip (Fig. 1.). The fluorescence appears to change, when allelopathically active cell of other plant species (cell-donor) interacts with acceptor cell (Roshchina and Melnikova, 1999).

3. MICROSPECTROFLUORIMETRY OF INTACT SECRETORY CELLS

Microspectrofluorimeters (detectors with optical probes of various diameters up to 2 mm) have been constructed in the Institute of Cell Biophysics, Russian Academy of Sciences (Karnaukhov et al., 1981, 1982;

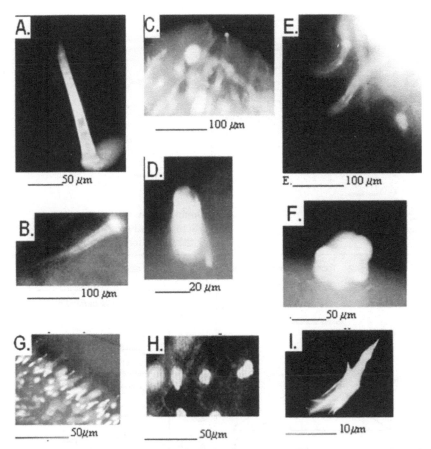

Fig. 1 The fluorescing images of secretory cells under luminescent microscope. A and B. Blue-fluorescing stinging and non-stinging secretory hairs of *Urtica dioica*, relatively on stem and leaf; C and D - green-yellow-fluorescing leaf glandular trichomes of *Lycopersicon esculentum* and *Solanum tuberosum*; E. - Blue-fluorescing leaf cells of *Achillea millefolium* ; F – yellow fluoresced gland of leaf *Calendula officinalis.*, G., H and I - secretory hairs, idioblasts and crystal on the surface on the root of *Ruta graveolens*, relatively.

1983; 1985; 1987). They can detect fluorescence from individual cells and even from a cell wall, large organelles and secretions in periplasmic space (space between plasmalemma and cell wall), as well as from the drops secreted by secretory cells and remaining on the cellular surface. Microspectrofluorimeters may record the fluorescent spectra or measure the intensities of the fluorescence at two separate wavelengths (Fig. 3). A special program "Microfluor" makes it possible to obtain the distribution histograms of the fluorescent intensities and to perform statistical analysis of the data using a Student t-test.

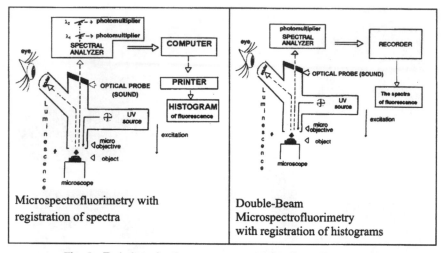

Fig. 2 Technique for the measurements of cell autofluorescence

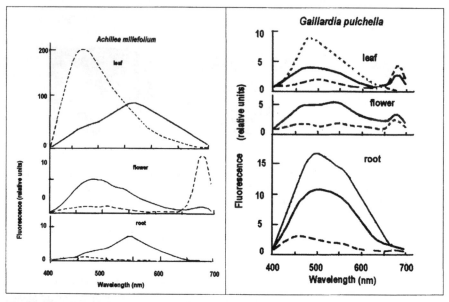

Fig. 3 The fluorescence spectra of secretory cells seen on flower, leaf and root of *Achilea millefolium* (left) and of *Gaillardia pulchella* (right). Unbroken fatty line – secretory hair; unbroken thick line – idioblast; broken line – non-secretory cell; dotted line – crystal on the surface (secretion).

Principle: The measurements of fluorescence spectra or /and fluorescence intensity by microspectrofluorimeters, which are a luminescent microscope combined with a multiplayer and a recorder (or a computer).

Material required: Microspectrofluorimetersm, object glasses (slices).

Procedure: The fluorescence spectra or the histograms of a fluorescence intensity of the objects are registered by microspectrofluorimeters (Karnaukhov *et al.*, 1981; 1982; 1983; 1985) as described earlier (Roshchina *et al.*, 1998; Roshchina and Melnikova, 1999; Roshchina, 2002; 2004; Roshchina *et al.*, 2002). Cells analysed were put on the slides (object glasses). The fluorescence spectra were recorded in the region between 400 and 700 nm and the values of the highest fluorescence within this region were also measured. In each treatment, 10 microspores were used for measuring the maximal fluorescence. The fluorescence intensity is measured in relative units.

Observations: Depending on the purpose, the fresh plant samples on slides may be observed immediately or after the histochemical staining with following analysis.

Statistical analysis: Statistical analysis consists of determination of Standard error of means (SEM) or the Student- t-testing. The measurement error was about 1-2% (10 spectra per one variant, n=10). Counting was performed in four or five replicates (the number of Petri dishes per a treatment). SEM for the fluorescence spectra was 2 %.

Experiment 2. *The analysis of the fluorescent spectra.*
Experiments with leaves and flowers of some plants, from family Asteraceae containing both sesquiterpene lactones and their derivatives such as azulenes, showed the increase of blue (420-480 nm) emission in flower of *Achilea*, while secretory cells in leaf and root also demonstrated green-yellow (500-550 nm) or yellow-orange (530-590 nm) fluorescence, peculiar to flavonoids quercetin or rutin (Fig. 4). But in *Gaillardia* species, secretory cell fluoresce, mainly, in green-yellow known for proazulenes gaillardine and others. In one of these intact cells during their development (Roshchina, 1999; Roshchina and Melnikova, 1999), for instance, such emission is observed only after the appearance of formed secretory cells on the petals of flower from medicinal plant *Achilea millefolium*.

Experiment 3. *Histograms of the fluorescence intensity.*
Using double-beam microspecrofluorimeter, the ratio of the fluorescence intensities red (maximum of emission at 640 nm) and green (maximum of emission at 530 nm) was measured. This ratio decreases in petals, which are covered with matured secretory cells, in comparison with petals of flower bud, where there are only few or no similar cells. This occurs due to a filling of mature secretory cells with green-fluorescing secretion. The histogram,

depicts the distribution of the fluorescing cells, shows the appearance of cells with increased green fluorescence, whereas, red fluorescence of chlorophyll becomes smaller due to the increase in secretory cells on the petal surface (Fig. 4).

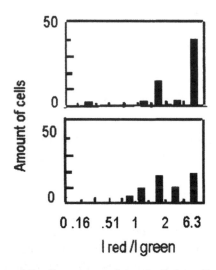

Fig. 4   The histograms of the fluorescence intensity (ratio red/green) for the secretory cells (the amount of cells fluorescing in green 520-540 nm) of developing leaf from *Achillea millefolium*. Upper histograms – leaf without developing secretory cells; lower histograms – after the appearance of secretory cells. (The amount of red fluorescing cells on the leaf surface decreased as can be seen from smaller heights of histograms).

Experiment 4. *The comparison of the fluorescence between various secretory cells in the same tissue.*

More information could be extracted from the observation of the fluorescence of roots from ruta [*Ruta graveolens* (family Rutaceae)] root, where idioblasts and secretory hair contains different secretory products as one can see, comparing their spectra. Quercetin and rutin as flavonoids found in the roots have less intensive emission than acridone type alkaloids such as rutacridone (Table 1). The intensivity of the rutacridone fluorescence in orange spectral region (590 -595 nm) is far more significant, than some of the flavonoids. The flavonoids include maxima in blue region of the spectra, but the flavonoid concentration was 10 times smaller, than fluoresced alkaloid. The fluorescence also depends on the medium (solvent). In oil and non-polar chloroform red-orange (580-600 nm) fluorescence is observed, but polar solvents and water shift the maxima to short wavelength regions, even from orange to blue. Thus, the study of the species show that allelochemicals

Table 1. The maxima (λ) in the fluorescence spectra of intact cells on the *Ruta graveolens* roots, extracts from the root tissue (1 hour of soaking) and known individual compounds of the tissue

Root intact cells	λ (nm)	Solvent for root extract (1:20)	λ (nm)	Pure substance	λ (nm)	Crystals dissolved (0.5 mg) in 0.5 ml of solvent
Non-secretory cell	480	water immersion oil	415, 456 / 450 / 590-600	Rutacridone	600-605	456,595(acetone), 480,585 (chloroform), 590-600 (immersion oil)
Idioblasts						
1.	475, 550 or 610	Purified mentholic oil	450 / 590-600	Rutacridone glycoside	595	435 (water) / 592 (ethanol)
2. (fluoresced in yellow and red-orange)						
Idioblasts (fluoresced in red-orange)	610	Purified mentholic oil + water	450 / 590	Rutacridone (acetone) + rutacridone glycoside (water)	-	456,585
Crystals on the surface with emission						
1. orange	610	Chloroform	480, 585	Quercetin	450, 610,	460 (acetone)
2. yellow	530					
3. green-yellow	490-520					
Tip of root (young root),	590-595	80 % acetone	430	Rutin	610	438, 600 (acetone)
Secretory hair (fluoresced in blue)	480	Water, then 80 % acetone	470			
Secretory hair (fluoresced in red-orange)	595 or 600					

concentrated in secretory cells may give an information about accumulation of the compounds. This permits to see it during the plant development or to estimate the state of the composition of the secretions with allelochemicals, which changed under many external factors (Roshchina, 2003; 2005). Using scales of microspectrofluorimeter, the investigator may know the accumulation of medicinal drugs - allelochemicals in the plant example. Root excretions from the tip could be compared with the content of idioblasts and secretory roots hairs. Host-killing allelochemical alkaloid rutacridone, in particular, is located in idioblasts and accumulated in secretory rutacridone-enriched tissues of the root tips. Rutacridone killed the microorganism and fungal parasite but can play role in allelopathic relations between plant species (Roshchina, 2002).

4. LASER SCANNING CONFOCAL MICROSCOPY OF INTACT SECRETORY CELLS

Principle: A confocal microscopy as a modification of luminescent microscope may produce images of high quality from fluorescing cells and permits the study of cells structures (see Chapter 8).

Materials required: Karl Zeiss Laser Scanning Confocal Microscope LSM 510 NLO "Carl Zeiss Karl Zeiss, object glasses, cover glasses

Procedure: The fluorescence of the cells are observed and measured on the object glasses (slides). All experiments were performed at room temperature 20-22 °C. The fluorescence image of the cells was also seen by the water immersion with laser scanning confocal microscope (LSCM) LSM510 NLO "Carl Zeiss Karl Zeiss". The excitation by three types of lasers: Argon/2 (λ 458, 488, 514 nm), HeNe1 (λ 543nm) and HeNe2 (λ 633 nm) were used for the emission (Roshchina *et al.*, 2004). In the experiments, three photomultipliers can catch the fluorescence, separately or simultaneously by use the pseudocolor effects. The image analysis was done with computer programmes LSM 510 and Lucida Analyse 5. At the excitation by lasers with wavelengths of 458, 488, 543 and 633 nm, the registration of the fluorescence was at 505-630 nm, 650-750 nm and 650-750 nm, respectively. Pseudocolours were according to the exciting wavelength blue for 488; green for 543 and red for 633. Summed image was seen when the images consisted of the pseudocolours were superimposed and mixed.

Experiment 5. *LSCM images of secretory hair*
Secretory hairs of allelopathically active species contain various secondary metabolites (allelochemicals) which may fluoresce in blue-green on the red

fluorescing (due to chlorophyll) surface of the leaf or stem tissue, as seen in Chapter 8. The emitted structure and the intensity of the interior are clearly seen.

5. CELLS STAINED WITH FLUORESCENT ALLELOCHEMICALS DYES

When allelochemicals and pure fluorescent substances (10^{-6}-10^{-5}M) were added to the acceptor cell, the changes in their fluorescence could be seen (Roshchina, 2004, 2005 a,b). This may be used to study the allelopathic mechanisms at cellular level.

Procedure: The fluorescence spectra of the ethanol and water solutions of the allelochemicals 10^{-5} –10^{-7} M were recorded with spectrofluorimeter Perkin –Elmer 550 in 1-cm cuvettes. The excitation wavelength was 360 nm. LSCM images were analysed as in section 9.4.

Experiment 7. *Analysis of the allelochemicals' fluorescence out cell and after the stained cellular models - unicellular microspores Equisetum arvense and Hippeastrum hybridum.*

Table 2 shows the possibility for various allelochemicals to be bound in various cellular compartments. Alkaloids, which have the anticholinesterase activity, such as physostigmine, berberine and sanguinarine, concentrated and fluoresced with orange colour on the surface of the cell. Pollen germination in artificial nutrient medium decreased after the addition of berberine and sanguinarine. The alkaloid rutacridone and sesquiterpene lactones (gaillardine and grosshemine), which inhibited the pollen germination passed through the plasmalemma into the cell and nucleus became green fluorescent (rutacridone) or blue lightening (sesquiterpene lactones). The substances also reacted with DNA or nucleic acid-protein complexes, inducing similar fluorescence.

Experiment 8. *Fluorescence of cells stained with rutacridone*
Allelochemical rutacridone penetrates the cell and binds with nucleic acids (Roshchina, 2002; 2005b). When these cells are excited with ultra-violet or violet light, they changed fluorescence. After similar histochemical staining, nucleus and chloroplasts as DNA- containing organelles fluoresce in green, unlike other cytoplasm, which looks as orange-fluoresced (see Chapter 8).
Fluorescent allelochemicals could be used as fluorescent dyes for the study of their interactions with acceptor cells in allelopathic mechanisms.

Table 2 Effects of allelochemicals on the fluorescence and germination of microspores

Allelochemical	Fluorescence (color/maxima in nm)			Cellular component target of fluorescent staining	Effect on the germination of microspores I_{50} or S_{50} (M)	
	Pure compound	In interaction with microspores				
		E. arvense	H. hybridum		E. arvense	H. hybridum
Sesquiterpene lactones						
Artemisinine	Blue/440	Blue-green /450, 530	Blue/440	Nucleus, chloroplasts	$I = 4 \times 10^{-4}$	$S = 10^{-5}$
Azulene	Blue/425	Blue/425, 550	Blue/425, 550	Nucleus, chloroplasts	N	N
Gaillardine	Blue/420	-	Blue-green /495-500	Nucleus	$I = 10^{-5}$	$I = 8 \times 10^{-6}$
Tauremizine	Blue/440	Blue/440	Blue/440	Nucleus	$I = 10^{-5}$	-
Alkaloids						
Atropine	Blue/415	Blue/465-470	Blue/465-470	ChR in Plm	$I = 10^{-5}$	N
Berberine	Yellow/540	Yellow/ 540-545	Yellow-orange/ 545-575	ChE ,Plm , CW	$I = 10^{-5}$	$I = 5 \times 10^{-7}$
Casuarine	Blue/475	Green/500	Blue/475	Nucleus, Glc	$I = 10^{-4}$	$I = 10^{-4}$
Chelerythrine	Green-yellow/ 510, 540	-	Green-yellow/ 510, 540	ChE in Plm, CW	$I = 5 \times 10^{-5}$	$I = 2 \times 10^{-5}$
Colchicine	Blue/440	Blue/475	Green/530	Tubulin in microtubules	$I = 10^{-7}$	$I = 7 \times 10^{-6}$
Glaucine	Blue/440	Blue-green/ 440, 510	Blue-green/ 440, 500	Nucleus, ChE, Plm, CW	Weak I	N
Physostigmine	Blue/410	N	N	N	N	N
Rutacridone	Orange/595	Green /510	Green /475, 510-550	Nucleus	N	N
Sanguinarine	Orange/595	Orange/575	Orange/595	ChE, Plm, CW	$I = 8 \times 10^{-7}$	$I = 5 \times 10^{-6}$
d-Tubocurarine	Blue/415	-	Blue/465-475	ChR, Plm	$I = 10^{-8}$	N
Yohimbine	Blue/455	-	Blue/475	AdR, Plm	$I = 5 \times 10^{-7}$	N

*Plm – plasmalemma, CW –cell wall, ChE- cholinesterase, ChR – cholinoreceptor, AdR –adrenoreceptor, Glc–glucosidase, I – inhibition, S-stimulation, I_{50} or S_{50} - concentration, which inhibits or stimulate the process by 50 %, N - no effect, (-) – no experiments

6. SUGGESTED READING

Karnaukhov, V.N., Yashin, V.A., Kulakov, V.I., Vershinin, V.M. and Dudarev, V.V. (1981). Technique to study the luminescent microobjects. DDR Patent N147002, 1-32.

Karnaukhov, V.N., Yashin, V.A., Kulakov, V.I., Vershinin, V.M. and Dudarev, V.V. (1982). Apparatus to investigate the fluorescence characteristics of microscopic objects. US Patent, No. 4, 354, 114, 1 - 14.

Karnaukhov, V.N., Yashin, V.A., Kulakov, V.I., Vershinin, V.M. and Dudarev, V.V. (1983). Apparatus to investigate the fluorescence characteristics of microscopic objects. Patent of England 2.039.03 R5R.CHI.

Karnaukhov V.N, Yashin V.A. and Krivenko V.G. (1985). Microspectrofluorimeters. *Proceedings of First Soviet -Germany International Symposum. Microscopy, Fluorimetry and Acoustic Microscopy.* Moscow, p.160-164.

Reigosa Roger, M.J. and Weiss, O. (2001). Fluorescence technique. In: *Handbook of Plant Ecophysiology Technique,* (Ed., M.J. Reigosa Roger) Pp. 155-171. Kluwer Academic Publishers, Dordrecht.

Roshchina, V.V. (2002). Rutacridone as a fluorescent dye for the study of pollen. *Journal of Fluorescence* **12**: 241-243.

Roshchina, V.V. (2003). Autofluorescence of plant secreting cells as a biosensor and bioindicator reaction. *Journal of Fluorescence* 13 : 403-420.

Roshchina, V.V. (2004). Cellular models to study the allelopathic mechanisms. *Allelopathy Journal* **13 : 3-16.**

Roshchina V.V. (2005a). Contractile proteins in chemical signal transduction in plant microspores. *Biological Bulletin, Ser. Biological.* **3** : 281-286.

Roshchina, V.V. (2005b). Allelochemicals as fluorescent markers, dyes and probes. *Allelopathy Journal* **16** : 31-46.

Roshchina, V.V. and Melnikova E.V. (1999). Microspectrofluorimetry of intact secreting cells with applications to the study of allelopathy. In : *Principles and Practices in Plant Ecology. Allelochemical Interactions.* (Eds., Inderjit, K.M.M. Dakshini, C.L. Foy). pp 99-126. CRC Press. Boca Raton, USA.

Roshchina, V.V., Melnikova, E.V., Mitkovskaya L.I. and Karnaukhov, V.N. (1998). Microspectrofluorimetry for the study of intact plant secretory cells. *Journal of General Biology (Russia).* **59**:531-554.

Roshchina, V.V. and Roshchina, V.D. (1993). *The Excretory Functions of Higher Plants.* Springer-Verlag, Berlin. 314 pp.

Roshchina, V.V., Yashin, V.A. and Kononov, A.V. (2004). Autofluorescence of plant microspores studied by confocal microscopy and microspectrofluorimetry. *Journal of Fluorescence* 14 : 745-750.

Wang, X.F. and Herman, B. (Eds.) (1996). *Fluorescence Imaging Spectroscopy and Microscopy.* John Wiley and Sons, London.

Weiss, O. and Reigosa Roger, M.J. (2001). Modulated fluorescence. In: *Handbook of Plant Ecophysiology Technique,* (Ed., M.J. Reigosa Roger) pp. 173-183. Kluwer Academic Publishers, Dordrecht.

Methods of Analytical Biochemistry and Biophysics

Biochemical Approach to Study Oxidative Damage in Plants Exposed to Allelochemical Stress: A Case Study

R. Cruz-Ortega* and A.L. Anaya

1. INTRODUCTION

Allelopathic interactions are mediated by secondary metabolites released from donor plants to the environment and have an influence on the growth and development in both natural and agro-ecosystems. These allelochemicals belong to many chemical groups and have different sites and modes of biochemical action. In general, when the effect of these allelochemicals decreases the growth of the receiver plant, it is considered as a biotic stress called *allelochemical stress'*. This environmental stress factor can act as a mechanism of interference and can influence the pattern of vegetation, weed growth and crop productivity (Dakshini *et al.* 1999; Weir *et al.* 2004; Romero-Romero *et al.* 2005). Allelochemicals can have several molecular targets in the receiver plants, interfering with different cellular processes and thus inhibiting plant growth. Previous studies have shown that allelochemical stress can cause an (i) oxidative damage [evidenced by an increase in reactive oxygen species (ROS)] to membranes and (ii) modification in some antioxidant enzyme activities (Cruz-Ortega *et al.* 2002; Bais *et al.* 2003; Weir *et al.* 2004).

Our research has focussed on understanding the allelochemical modes of action of the weed *Sicyos deppei* G Don (Cucurbitaceae). *S. deppei* is a very

Departamento de Ecología Funcional. Instituto de Ecología, Universidad Nacional Autónoma de México, México D.F. 04510. *E-mail: rcruz@miranda.ecologia.unam.mx harmful weed and is endemic in the central part of Mexico. It causes an allelochemical stress in crop plants such as tomato (*Lycopersicon esculentum*); an aqueous leachate of this weed inhibits 70% tomato radicle growth (Romero-Romero *et al.* 2002; 2005). To relate if, the inhibitory effect is due to an oxidative stress caused by the allelochemical stress produced by *S. deppei*, we have measured the following: (i) activities of antioxidant enzymes: catalase (CAT), ascorbate peroxidase (APX), glutathione reductase (GR) and superoxide dismutase (SOD); (ii) levels of lipid peroxidation and (iii) levels of reactive oxygen species (ROS). In this chapter, we will describe the biochemical methodologies that we use to test, if allelochemical stress causes an oxidative damage in plants, using as an example the case study of *S. deppei* as the allelopathic plant and tomato as the receiver or damaged crop.

2. BIOASSAYS

Experiment 1. *Bioassays with allelochemical aqueous leachate.*

Principle: Phytotoxic effects of the aqueous leachate of an allelopathic plant can be tested *in vitro* bioassays. Test or target plants are placed in contact with 0.5 % aqueous leachate from the allelopathic plant. Germination and radicle growth can be monitored during time-course experiments (i.e. after 24, 48 and 72 h of treatment), but in this chapter we will include only the results obtained after 72 h of treatment.

Material required: Air-dried (27–30°C) aerial parts of *S. deppei* to prepare aqueous leachate, Seeds of tomato (*Lycopersicon esculentum* Mill. cv. Río Grande), Osmometer, Growth Chamber Laminar flow hood.

Procedure: Allelopathic aqueous leachate is prepared by soaking dried leaves (1g/100 mL or 1% w/v) in distilled water for 3 h. This leachate is filtered through Whatman paper (No. 4) and then through a sterile Millipore membrane (0.45 mm). Then, it is poured in Petri dishes and mixed with agar (2%) for a final aqueous leachate concentration of 0.5%. The volume will depend on the size of the Petri dish, 3 mL of leachate plus 3 mL of agar are enough for a 6 cm Petri dish. Osmotic potential of the leachate is measured with a freezing-point osmometer (Osmette A, Precision System Inc.).

Bioassays are performed under sterile conditions in a laminar flow hood. Tomato seeds are previously washed and disinfected with 1% sodium hypochlorite. Seeds are germinated in the Petri dishes containing the *S. deppei* aqueous leachate. For control, seeds are germinated in 1% agar. Twelve seeds are placed on each Petri dish and kept in the dark at 27°C in a growth chamber. For enzyme activities, 40-50 Petri dishes are used per treatment. Primary roots (radicles) are excised after 72 h, frozen in liquid nitrogen and kept at -70 °C until use. For root growth response, experiments are conducted in a complete randomized design with four replicates. To determine the percentage of inhibition, data on primary root lengths are analyzed by an ANOVA test, using a Statistical Program Duncan's Multiple Range Tests (Mead *et al.* 2002).

Observations: Germination and radicle growth can be determined at 12, 24, 48 and 72 h and the inhibitory effect of the aqueous leachate can be calculated comparing root length of treated seedlings vs. data from control ones. In this case, we only used the 72 h treatment (Figure 1).

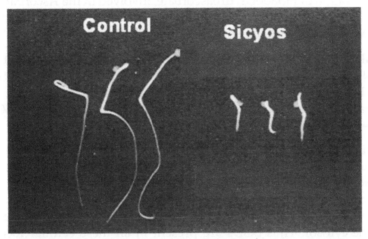

Fig. 1 Effects of aqueous leachate of *S. deppei* on radicle growth of tomato after 72 h of treatment.

3. ACTIVITY OF ANTIOXIDANT ENZYMES

Experiment 2. *Activity of antioxidant enzymes.*

Principle: One way to determine if the allelochemical stress is causing an imbalance in Reactive Oxygen Species (ROS) is measuring the activity of the antioxidant enzymes. Their performance will be indicative, if the treated cells are getting rid of the level of ROS that are produced during the stress. Total protein extracted under native conditions is used to measure activity of enzymes with their proper substrate. Spectrophotometric assays are used to determined substrate consumption, or oxidation/reduction of co-factors to relate the activity of the enzymes. Catalase, ascorbate peroxidase, glutathione reductase and superoxide dismutase are part of the antioxidant system that plant cells have to control levels of free radicals (ROS). A modification in these enzyme activities can produce an imbalance on ROS and therefore, produce damage at the membrane level, in proteins, DNA, or in signal transduction mechanisms (Vranova *et al.* 2002).

I. Catalase (EC 1.11.1.6): Catalase enzyme protects the cell from the damage caused by hydrogen peroxide (H_2O_2), which is produced during the redox reactions under normal conditions or by an environmental stress. Control of H_2O_2 levels is complex and knowledge of mechanisms generating and relieving H_2O_2 stress is difficult, particularly in intact plants.

II. Ascorbate Peroxidase (EC 1.11.1.11): APX isoenzyme, plays an important role in the metabolism of H_2O_2 in higher plants. APX utilizes Ascorbate (Asx) as its specific electron donor to reduce H_2O_2 to water with the generation of monodehydroascorbate, involved in the Ascorbate-GSH cycle. Thus, APX in combination with the effective Asx-GSH cycle functions to prevent the accumulation of toxic levels of H_2O_2.

III. Glutathione reductase (EC 1.6.4.2): It is a flavoprotein that catalyzes the NADPH-dependent reduction of oxidized glutathione (GSSG) to glutathione (GSH). This enzyme is essential for the GSH redox cycle which maintains adequate levels of reduced cellular GSH. A high GSH/GSSG ratio is essential for protection against oxidative stress.

IV. Superoxide dismutase (EC 1.15.1.1): Within a cell the superoxide dismutases (SODs) constitute the first line of defense against ROS. Superoxide radical (O_2^-) is produced where an electron transport chain is present, as in mitochondria and chloroplasts, but O_2 activation may occur in other subcellular locations such as glyoxysomes, peroxisomes, apoplast and the cytosol. Thus SODs are present in all these cellular locations, converting superoxide into hydrogen peroxide and water (i.e. copper/zinc SODs are typically found in the nuclei and cytosol of eukaryotic cells).

Material required: For total protein extract and enzyme activities: extraction buffer: 50 mM potassium phosphate, pH 7.0, 1 mM EDTA, 1% of PVP (100 mg/mL), for CAT, GR, and SOD and with 10 mM ascorbato for APX activity. Bio-Rad protein assay, centrifuge, spectrophotomer and quartz cuvette.

Protein content: The amount of protein in each extract can be determined by the Bradford method (Bradford, 1976), using BSA as a standard. Briefly, make a standard curve with 0, 2, 4, 6, 8, 10, 15 and 20 µg /mL BSA and mixed with 1 mL of Bio-Rad protein assay (diluted 1:4). Read standard curve and samples at A_{595} in a spectrophotometer, using as blank 1 mL of diluted Bio-Rad protein assay.

Enzyme assays: For enzyme activities, total protein is extracted from seedlings of controls and after 72 h-treatment under native conditions. Primary roots are homogenized in extraction buffer (for formulation see above). The homogenate is centrifuged at 3 000 g for 10 min at 4 °C to remove cellular debris.

Catalase (EC 1.11.1.6) activity: It can be estimated by determining the consumption of H_2O_2 (extinction coefficient 39.4 mM^{-1} cm^{-1}) at 240 nm for 3 min. 50-100 µg of total protein is used in each reaction. In a 1mL quartz cuvette, add the reaction medium that contains 50 mM potassium phosphate buffer (pH 7.0) (volume required to bring up to 1mL with protein), 10 mM H_2O_2, volume for 50 µg of protein extract. Using as a blank the reaction buffer, the activity is determined measuring at 240 nm the consumption of H_2O_2 for 3 min. Make at least 3-5 replicates of each measurement. The slope is determined as Absorbance/time and activity is calculated using the extinction coefficient of catalase, as follows (Aebi, 1974).

activity $= [m/\varepsilon \times mg \, protien \times mL \, reaction \, assay] \times 60$

Where,

m = slope from the curve absorbance/time

ε = molar extinction coefficient = 39.4 mM^{-1} cm^{-1}

Ascorbate peroxidase (EC 1.11.1.11) activity: It is estimated by determining the decrease in A_{290} [Extinction coefficient (ε) 2.8 mM^{-1} cm^{-1}] for 1 min. In a 1 mL quartz cuvette add the reaction medium containing 50 mM potassium phosphate buffer (pH 7.0) (the volume required to bring up to 1mL with the protein), 0.5 µM ASC, 0.1 µM H_2O_2 and volume required for 50 µg of total protein extract. Correction is done for the low, non-enzymatic oxidation of ASC by H_2O_2 (Jiang and Zhang, 2002). The slope is determined as Absorbance/time, and the activity is calculated using the extinction coefficient of APX, as follows:

activity $= m/\varepsilon \times mg \, protien \times mL \, reaction \, assay$

Where,

m = slope from the curve absorbance / time (1 minute)

ε = molar extinction coefficient = 2.8 mM^{-1} cm^{-1}

Glutathione reductase (EC 1.6.4.2) activity: It is determined by estimating the oxidation of NADPH at 340 nm (extinction coefficient 6.2 mM^{-1} cm^{-1}) for 3 min. In 1 mL quartz cuvette, add the reaction medium that contains 50 mM potassium phosphate buffer (pH 7.8), 2 mM Na_2EDTA, 0.15 mM NADPH, 0.5 mM GSSG and the volume for 50 µg of total protein extract. The reaction is initiated by adding fresh 0.15 mM NADPH. Corrections are made for the background absorbance at 340 nm, without NADPH (Jiang and Zhang, 2002). The slope is determined from the curve Absorbance/time, and the activity is calculated using the extinction coefficient of GR, as follows:

$$\text{activity} = [m/\varepsilon \times \text{mg protien} \times \text{mL reaction assay}] \times 60$$

Where,

m = slope from the curve absorbance / time

ε = molar extinction coefficient = 6.2 mM^{-1} cm^{-1}

Superoxide Dismutase (EC 1.15.1.1) activity: It is assayed by following the method of autoxidation of epinephrine (adrenochrome) described by Misra and Fridovich (1972). Briefly, 250 mg of frozen tissue is homogenized with 1 mL of 100 mM potassium phosphate buffer pH 7.8 and 0.1 mM EDTA. Then this homogenate is centrifuged at 6000 x g for 10 min at 4 °C. For activity determination, in 0.5 mL quartz cuvette add the reaction medium that contains: 325 µL of 50 mM of sodium carbonate buffer pH 10.2, 100 µL 0.5 mM EDTA, 0.50 µL total protein extraction and 25 µL of 10 mg/mL epinheprine (dissolved in 10 mM HCl, pH 2). Autoxidation of epinephrine is determined at 480 nm every 10 sec for 3 min, considering only readings after lag phase (2-3 min, until highest absorbance value starts to decrease). The activity is reported as µmol adenocromo/ min/ mg protein.

$$\text{activity} = [m/\varepsilon \times \text{mg protien} \times \text{mL reaction assay}] \times 60$$

Where,

m = slope from the curve absorbance / time

ε = molar extinction coefficient = 4020 M^{-1}cm^{-1}

Table 1   Antioxidant enzymes activities after 72 h treatment

	CAT	APX	GR	SOD
Control	21.968 ± 0.31	49.62±6.89	0.013±0.0028	9.21±0.046
Sicyos	32.587 ± 0.67	21.43±2.28	0.013±0.0015	6.93±0.065

Activities for CAT =Catalase µmol/min/mg protein; APX = Ascorbate peroxidise µmol/min/mg protein (E^{-5}); GR = Glutathione reductase nmol/min/mg protein; SOD = Superoxide dimutase mmol/min/mg protein (E^{-5})

Observations: Table 1 shows the activity of the antioxidant enzymes of tomato roots after 72 h of exposure of allelochemical stress caused by *S. deppei*. Catalase (CAT) activity increases by 1.5 fold; Ascorbate Peroxidase (APX) decreases 2.3 fold; Glutathione reductase (GR) activity does not change with the treatment; and Superoxide dismutase (SOD) decreases 1.3 fold.

4. DETERMINATION OF FREE RADICALS, H$_2$O$_2$ AND LIPID PEROXIDE

Experiment 3. *Determination of free radicals, H$_2$O$_2$ and lipid peroxidation.*

Principle: The production of free radicals is measured with luminol (5-amino-2, 3-dihydrophthalazine- 1, 4-dione), that is employed to amplify the chemiluminescence signals. Luminol is oxidized by several oxygen free radicals (O^{2-}, HO^-, $_1O^2$, H_2O_2) to an electronically excited aminophthalate anion that, upon relaxation to the singlet ground state, emits photons (Cadenas and Sies 1984). H_2O_2 can be determined by colorimetric assay, using titanium sulfate. Membrane lipid peroxidation is determined by measuring conjugated dienes, a product of lipid peroxidation.

Procedure

   (i) **Free radicals:** Total protein extract is obtained by homogenizing frozen tissue with the extraction buffer containing 50 mM phosphate buffer, pH 7.0 and 0.1 mM EDTA and then protein content is determined by Bradford method, as described earlier. In a scintillation vial, add 4 mL of the reaction medium containing 0.5 M ammonium acetate, pH 10.5, 0.25 mM sodium carbonate, 50 µM luminol; 250 µM CoCl, and 0.06 mg of total protein extract. Emittance is measured in a scintillation counter (Beckman LS6000SE, with a chemiluminescence programme). An initial reading of cpm with no sample is taken and a second one is taken after the 0.060 mg of protein extract is added. At least 10 readings per sample are taken and the highest value of cpm is used as a final value. The number of free radicals is calculated as follows:

[Final value of cpm – Initial value of cpm] 0.06 mg protein

  (ii) **Membrane lipid peroxidation:** To determine if free radicals cause damage to the membrane, conjugated dienes can be measured as a product of membrane lipid peroxidation, they are measured in the ultraviolet range (233 nm) as per the method of Recknagel and Glende (1984). Total protein is obtained from frozen tissue by homogenizing with extraction buffer as described for enzyme activities. After protein determination, 0.5 mg of protein from either control or treatment is dissolved in 1 mL of H_2O and 4 mL of a mixture of chloroform-methanol (2:1 v/v). This mixture is placed on ice for 30 min and then centrifuged at 260 g for 5 min. The lower chloroform layer is removed and transferred to a clean tube and placed in a water bath at 60 °C, to remove all the chloroform. The extracted chloroform-free lipids are dissolved in 1.6 mL cyclohexane (pure) and optical density of conjugated dienes is recorded at 230 nm, using a quartz cuvette and cyclohexane as a blank.

 (iii) **H_2O_2:** Its content can be estimated colorimetrically as described by Jana and Choudhuri (1981). H_2O_2 is extracted by homogenizing 50 mg of frozen tissue with 0.75 mL of 50 mM phosphate buffer (pH 6.5). This homogenate is centrifuged at 6000 x g for 25 min and then 0.25 mL of this extracted solution is mixed with 0.25 mL of 0.1% titanium sulfate in 20% (v/v) H_2SO_4. The mixture is centrifuged at 6000 x g for

15 min and the intensity of the yellow colour is measured at 410 nm, using 1.0 mL quartz cuvette and as a blank 3mL of 50 mM phosphate buffer plus 1 mL of 0.1% titanium sulfate in 20% (v/v) H_2SO_4. Hydrogen peroxide level is calculated by the extinction coefficient of 0.28 $\mu mol^{-1} cm^{-1}$ as follows.

$$H_2O_2 = \frac{A410}{\varepsilon \times 0.05 \text{ mg tissue}}$$

where,

ε = extinction coefficient = 0.28 $\mu mol^{-1} cm^{-1}$

Observations: Table 2 shows the levels of free radicals determined by luminol chemiluminescence, H_2O_2 levels, and membrane lipid peroxidation determined by conjugated dienes. Total free radicals significantly increased by 1.4 fold, hydrogen peroxide significantly decreased 1.3 fold, and lipid peroxidation increased significantly 1.5 fold.

Table 2 Effects of *S. deppei* on free radicals, H_2O_2 and lipid peroxidation of *L. esculentum* roots at 72 h after treatment.

	Free radicals (cpm/mg)	H_2O_2 (μmol)/mg tissue	Lipid peroxidation (ΔA_{233})
Control	46400 ± 5300 a	66.8 ±2.1a	0.183 ± 0.010a
S. deppei	66700 ± 9900 b	48.6 ± 4.1b	0.270 ± 0.034b

Different letters mean statistical differences (Duncan test, P<0.001)

5. VISUALIZATION OF REACTIVE OXYGEN SPECIES (ROS)

Experiment 4. *Visualization of ROS with Confocal Microscopy.*

Principle: ROS production can be monitored by imaging the ROS-sensitive fluorescent dye dichlorofluorescein (DCF) in a confocal microscope.

Material required: Stock solution of 25 mM H_2DCFDA, carboxy-2',7'-dichlorofluorescein diacetate (Sigma Co) dissolved in dimethyl sulphoxide (DMSO). Confocal Microscope BIORAD 1024, Laser KrAr with inverted microscope Nikon TMD 300.

Procedure: From bioassays, control and treated seedlings (seed and roots at 48 and 72 h of treatment) are stained for 10-15 min with 25 μM DCFDA in distilled water. Then fluorescence intensity of the dye is observed using the confocal microscope BIORAD 1024 (488nm dichroic and 510-560 nm emission). DCFDA fluorescence increases as the dye is oxidized by ROS to dichlorofluorescein (DCF).

Observations: Figure 2 shows confocal images of 48 h tomato roots exposed to *S. deppei* aqueous leachate and then stained with DCFDA. A higher fluorescence is observed in treated-root hairs.

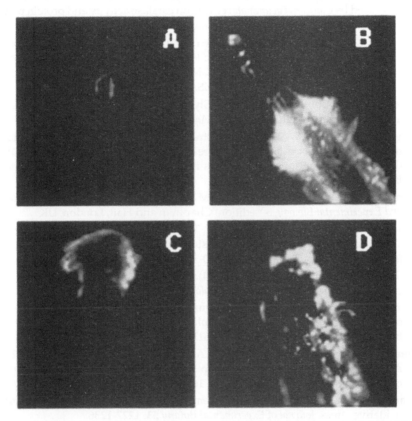

Fig. 2 Confocal imaging. *L. esculentum* roots after 48 h of treatment, and with 25 μM DCF for 10 min. Control roots (A and C), *S. deppei* treated-roots (B and D). Panels C and D are higher magnification.

6. REFERENCES

Aebi, H. (1974). Catalase. In: *Methods of Enzymatic Analysis*. H.U. Bergmeyer (ed), pp. 673-684. Verlag Chemie, Weinheim, Germany.

Bais, H.P., Vepachedu, R., Gilroy, S., Callaway, R.M. and Vivanco, J.M. (2003). Allelopathy and exotic plant invasion: From molecules and genes to species interactions. *Science* **301:** 1377-1380.

Bradford, M.R. (1976). A rapid and sensitive method for the quantitation of microgram quantities of protein utilizing the principle of protein-dye binding. *Analytical Biochemistry* **72:** 248-254.

Cadenas, F. and Sies, H. (1984). Low-level chemiluminescence as an indicator of singlet molecular oxygen in biological systems. *Methods in Enzymology* **105**: 221 231.

Cruz-Ortega, R., Ayala-Cordero, G. and Anaya, A.L. (2002). Allelochemical stress produced by *Callicarpa acuminata*: effects on catalase activity and protein pattern synthesis of crop plants. *Physiologia Plantarum* **116**: 20-27

Dakshini, K.M.M., Foy, C.L. and Inderjit. (1999). Allelopathy: one component in a multifaceted approach to Ecology. In: *Principles and Procedures in Plant Ecology: Allelochemicals Interactions*. Inderjit, K.M.M. Dakshini and C.L.Foy (eds), pp. 3-14. CRC Press, Boca Raton, Florida, USA

Jana, S. and Choudhuri, M. A. (1981). Glycolate metabolism of three submerged aquatic angiosperms during ageing. *Aquatic Botany* **12**: 345-354.

Jiang, M. and Zhang, J. (2002). Water stress-induced abscisic acid accumulation triggers the increased generation of reactive oxygen species and up-regulates the activities of antioxidant enzymes in maize leaves. *Journal of Experimental Botany* **53**: 2401-2410.

Mead, R., Curnow, R.N. and Hasted, A.M. (2002). *Statistical Methods in Agriculture and Experimental Biology.* 3rd edition. Chapman and Hall, London, UK

Misra, H.P. and Fridovich, I. (1972). The role of superoxide anion in the autoxidation of epinephrine and a simple assay for superoxide dismutase. *Journal of Biological Chemistry* **247**: 3170-3175.

Recknagel, R.O. and Glende, E.A. (1984). Spectrophotometric detection of lipid conjugated dienes. *Methods in Enzymology* **105**: 331-337.

Romero-Romero, T., Anaya, A.L. and Cruz-Ortega, R. (2002). Screening for effects of phytochemical variability on cytoplasmic protein synthesis pattern of crop plants. *Journal of Chemical Ecology* **28**: 617-629.

Romero-Romero, T., Sánchez-Nieto, S., Anaya, A.L. and Cruz-Ortega, R. (2005). Allelochemical and water stress in roots of *Lycopersicon esculentum*: a comparative study. *Plant Science* **168**: 1059-1066.

Vranová, E., Inzé, D. and Van Breusegem, F. (2002). Signal transduction during oxidative stress. *Journal of Experimental Botany* **53**: 1227-1236.

Weir, T.L., Park, S.W. and Vivanco, J.M. (2004). Biochemical and physiological mechanisms mediated by allelochemicals. *Current Opinion in Plant Biology* **7**: 472-479.

Chapter

Cholinesterase Activity as a Biosensor Reaction for Natural Allelochemicals: Pesticides and Pharmaceuticals

A.Y. Budantsev* and V.V.Roshchina**

1. INTRODUCTION

Cholinesterase of living cells is a sensitive enzyme to artificial and natural toxins (Augustinsson et al., 1963; Budantsev, 2005; Budantsev and Roshchina, 2005). The enzyme similar to sensitive cholinesterase of animals, is found in plants and microorganisms (Roshchina, 2001). Moreover, it is also used a bioreactor for determination of pesticides (Botre et al., 1994). Many allelochemicals, especially pharmaceutical alkaloids, demonstrate the pesticidal characteristics, affecting cholinesterases of plants and animals (Narahashi, 1979; Ulrichova et al., 1983; Atta-ur-Rahman et al., 2001; Roshchina, 2004). The phenomenon is also of interest to understand the allelopathic mechanisms of growth regulation, in particular the acetylcholinesterase stimulates the growth reactions of animals (Sternfeld et al., 1998). These effects may be analyzed using the biotests based on the preparation from either animal or plant cells as well as from pure commercial enzyme cholinesterase (Evtugin et al., 1999). Biotests based on animal and plants cholinesterases to determine alkaloids (known as allelochemicals), pharmaceuticals and pesticides are given in this chapter.

*Institute of Theoretical and Experimental Biophysics, Russian Academy of Sciences. Institutskaya str., 3, Pushchino, Moscow region, 142290, Russia. E-mail: budantsev@mail.ru.
**Institute of Cell Biophysics, Russian Academy of Sciences. Institutskaya str., 3, Pushchino, Moscow region, 142290, Russia. E-mail: roshchina@icb.psn.ru

2. PREPARATION OF CHOLINESTERASE-CONTAINING MATERIAL

The cholinesterase to determine the toxic activity may be chosen: (i) in pure form of commercial enzyme from animals in a water buffer solution or using biosensors, enzyme preparation impregnated into a rigid matrix that significantly activates the enzymic activity and (ii) in the form of crude extracts from plant or animal tissues.

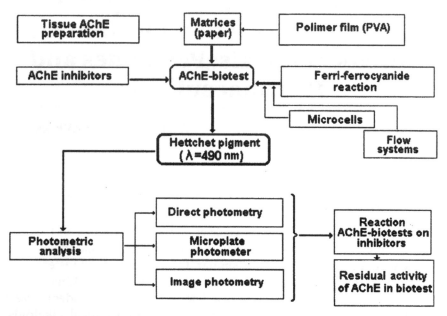

Fig. 1   The main scheme of AChE biotests' technology (AChE – acetylcholinesterase; PVA – polyvinil alcohol)

2.1   Paper Biotests of Animal Tissue Cholinesterase Preparations

Recent technology of manufacturing and biotesting includes the special biosensors. Biosensors are sensitive tests consisting of paper matrixes and tissue enzyme preparations (Budantsev and Litvinova, 1996; Budantsev *et al.*, 1997). One mode of their application is production of sensitive paper biosensors based on the acetylcholinesterase (AChE) isolated from animal tissues (Budantsev *et al.*, 1998). Among the biosensor systems are the enzyme tissue preparations viz., tissue sections, tissue homogenates, fractions of cells and intracellular organelles. The advantages of using matrixes for biotests are : (i) preservation of biotest activity for 6 mon (stored at room temperature), (ii) convenience in using control tests containing Congo red linear dependence concentration - optical density of dye in range of 125.0-250.0 – 3.9-7.8 mcg/ml (microplate photometer and (iii) high sensitivity AChE-biotest (on eserine) - 10^{-5} - 10^{-6} M.

Principle: Biosensors consist of paper matrixes and tissue enzyme preparations, often the pure enzyme AChE or AChE-containing cells. As seen in Fig.1, main scheme of the preparation technology and procedure includes : (i) the preparation of same kinds of biotests-biosensors, which are paper matrixes impregnated with tissue preparation of AChE and covered by polymer film.; (ii) biochemical reactions of the AChE activity with and without inhibitors tested and (iii) the photometric analysis of the samples for quantitative estimation of the biochemical reactions.

Materials required

Matrixes: Three types of paper filters (i). Filtrak, Germany, the filter Number 388 [(0.025) - soft, wide pores]; filters No. 90 (0.15) and No. 90 (0.25) - dense, narrow pores); (ii). chromatographic papers FN-5, FN-11 (Germany); the kapron membrane (pore size 0.2 microns) and (iii). the membrane 'Vladipor' MFA-MA No. 6 (Kazan industrial associations 'Tasma')] were applied to matrixes for biotests. The circles (6 - 10 mm dia) are cut out from the matrixes materials by special instrument. The most stable results are received with use of matrixes of chromatographic papers FN-5, FN-11.

Enzyme: Commercial acetylcholinesterase preparation - electrical organ acetone powder (extract) from electric eel (*Electrophorus electricus*) (Sigma, E2384) was used.

Procedure: Some details of AChE-biotests preparation are shown in Fig. 2.

The preparation of AChE-biotest includes the following stages. (i). the preparation of homogenate from tissue with high cholinesterase activity (electrical organ acetone powder). Homogenate are prepared in phosphate

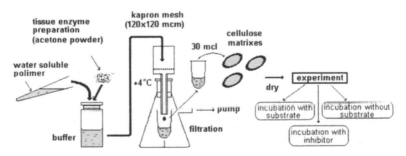

Fig. 2 The scheme of preparation of AChE-biotests (details in text)

buffer, distilled water or 2 % water solution of polyvinyl alcohol (PVA, m.m.49000, Fluka). The concentration of actone powder is 8 mg/ml. (ii). 30 mcl of the homogenate put up paper matrix (the diameter – 7-8 mm) and (iii). the matrices are dried at room temperature. The AChE activity in biotests is determined by ferrycyanide/ ferrocyanide reaction (Karnovsky and Roots, 1964). Aliquot of the homogenate is used for the biochemical analysis of AChE activity by the Ellman method (Gorun *et al.*, 1978) as described in section 3.2.

2.2   Extracts From Animal and Plant Material

Principle: The use of water extracts from a human or animal tissue and from plant tissues.

Material required: Horse (blood) serum preparation, human (blood) serum preparation, vegetative microspores of horsetail (*Equisetum arvense* L.) collected from meadows in April and May, generative microspores (pollen, male cells) or anthers of knight's star (*Hippeastrum hybridum* L.) grown in a green-house, the anthers from flowers of Liliales sp. plants were chosen. *Lilium tenuifoliwn* Pisch. (Liliaceae) with red flowers and *Gladiolus hybridus* with white flowers (Iridaceae) were grown outdoors in the field. Knight's star *(Hippeastrwn sp.)* and kafir-lily (*Clivia sp.*) hybrids (both from the Amaryllidaceae family) grown indoors, or cultivated cellular root tissue of ruta (*Ruta graviolens* L.) (The anthers of lily and gladiolus flowers were kept in a hermetically sealed vial at -24°C for 6 mon. Anthers from knight's star and kafir-lily were used immediately or kept in a hermetically sealed vial at 4°C for two wk; this did not influence the ChE activity).

Reagents: 20 mM Na-K-phosphate buffer pH 7.4

Procedure: Water solutions (in 20 mM Na-K-phosphate buffer pH 7.4) of liophilized commercial preparation from animal blood serum (5-6 mg/ml) or the extracts from vegetative microspores of horsetail (*Equisetum arvense*) or *Hippeastrum hybridum* (pollen grain) microspores (150 mg of microspores in 30 ml for 1 h) were used. If anthers were chosen for the analysis, just before the experiments, the anthers were homogenized in 20 mM Na-K-phosphate buffer pH 7.4 with an electric homogenizer using a teflon-glass system at 0°C for 60 sec and then filtered through nylon gauze.

3. BIOCHEMICAL ASSAYS OF ENZYME ACTIVITY

3.1   Formation of Red-Brown Product in Ferricyanide/Ferrocyanide Reaction (Karnovski and Roots Method)

Principle: The ferricyanide/ferrocyanide reaction is known as reaction Karnovsky-Roots (Karnovsky and Roots, 1964). A final product of this reaction is insoluble red-brown Hettchet's pigment (copper ferrocyanide, maximum of absorption at λ = 480 HM (Wenk *et al.*, 1973).

Materials required: AChE preparations, commercial acetylcholinesterase, horse (blood) serum preparation, human (blood) serum preparation, or matrix biotests with these immobilized enzyme (see section 15.2.1)

Reagents: Acetylthiocholine iodide or chloride (ATCh, Sigma); maleate buffer 0.1 M, pH = 6.0; sodium citrate; $CuSO_4 \cdot 5H_2O$; distilled H_2O; potassium ferricyanide; commercial acetylcholinesterase or water extract any cholinesterase-containing animal or plant.

Procedure: Cholinesterase activity in analyzed tissue or the matrix (biotest with immobilized AChE) is determined in the incubation media [consisting of substrate ATCh - 34 mmol; maleate buffer 0.1 M, pH = 6.0 – 6.5 ml; sodium citrate 0.1 M – 0.5 ml; $CuSO_4$ $5H_2O$ 0.03M – 1.0 ml; distilled H_2O (or inhibitor in variant with toxin analyzed) – 1.0 ml; potassium ferricyanide 0.005 M - 1 ml.] Volume of incubation media in one test - 400 mcl. As a blank (control sample), a treatment of the exposure without the substrate is used. If inhibitory effects of allelochemical (or any toxin) are analyzed, before the substrate addition the sample was preliminary exposed to allelochemical inhibitor. Two methods for the AChE-biotests may be recommended: (i) in microcells ('stationary conditions') and (ii) in flowing columns-reactors ('dynamic conditions').

I. The incubation of AChE-biotests in the microcells
In cells of 0.5 volume ml at the special incubation plate, the biotests (on one in a cell) are located. In each cell, 400 mcl of the incubation medium is added (see Section 3.1) air is squeezed out with thin glass stick from under the biotest and the incubation begins at room temperature. After the incubation each cell with the biotest is washed by distilled water (3 x 200 ml) and the reaction is stopped by adding 96 % ethyl alcohol (100 ml per cell). After 15 min, alcohol is removed from the cells and the biotests are dried at room temperature for 5-10 h (or at 50°C for 1-2 h).

II. The incubation of AChE-biotests in the flowing columns-reactors
AChE-biotests are located in a special column-reactor connected with the peristaltic pump and the incubation is done in flowing conditions. For this purpose special flowing, termistatic column-reactor is used (described in Supplement 2).

Observations: The hydrolysis of acetylthiocholine during 1 h leads to the formation of red-brown coloured product, which can be seen without any technique (see Experiment 1) or measured by photometric technique in spectral region 480–490 nm (see Experiment 2). Solution containing inhibitor is added into a media or on a AChE-biotests before incubation. In the last case, AChE-biotests are exposed for 30 min or 1 h and then incubated in the medium with substrate.

Experiment 1. *Visual observations of reaction*

The reactions of AChE-biotests results in the red-brown product (Hettchet pigment). The reaction of AChE-biotests on inhibitors can be estimated visually. The residual activity of AChE in biotests after the action of different concentrations of eserine (physostigmine) and proserine (neostigmine) is seen in Fig. 3.

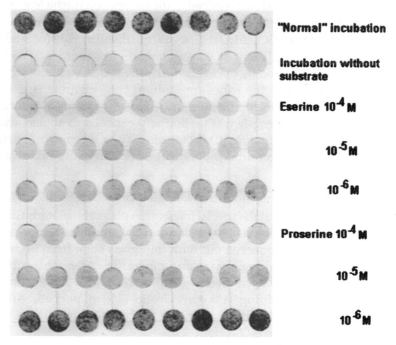

Fig. 3   AchE-biotests after normal colour reaction and after action of inhibitors

Experiment 2. *Measurements of optical density*

The quantitative estimation of biotests reaction can be expressed in terms of optical density (D) or in terms of factor of reflection coefficient (K_{ref}):

$$D = \lg I_0 / \lg I_e,$$

where I_0 - intensity of light passed through a pure (clean) matrix of the biotest, I_e - intensity of light after passing through the biotest after analytical procedure;

$$K_{ref} = R_e / R_0,$$

where R_0 - light intensity of the biotest, reflected from a pure (clean) matrix, R_e - light intensity reflected from the biotest after analytical procedure. All measurments are made at $\lambda = 480$ (490 nm, maximum absorption of Hettchet pigment)

In case of measurements D with the help TV-computer analysis, meanings I_0 and I_e are replaced with the appropriate meanings S_0 and S_e of a grayscale (256 gradation at 8-bit images).

The significance of biotests D, measured in the air or clean matrixes is used in the microplate photometers.

In the photometric analysis, a calibration of samples is made with the 5-30 ml of the water solutions of Congo red dye (Congo red Ind., Beijing Huagonchang) pH 5.6. This dye has the absorbance maximum 500 nm. Before the measurements, the tests are dried up on air at 50°C for 60 min. The absorbance of Congo red is also used as a blank for the absorbance measurements of AchE-biotests on the matrixes.

Experiment 3. *Estimation of residual enzyme activity in the biotest*
To calculate the residual activity of AChE-biotest after AChE inhibitors action on the biotest, three measurements will be done : (i) measurement D_n of the biotests after incubation with substrate, (ii) measurement $D_{w/s}$ after incubation without substrate and (iii) measurement D_{ing} after incubation with substrate and inhibitor. The amount of residual activity (A) will be calculated in percentage with the following formula:

$$A = [(D_{ing} - D_{w/s}) / (D_n - D_{w/s})] \times 100 \ (\%)$$

For example, Table 1 shows the results of the calculation D (Experiment 2) and residual activity A (Experiment 3) in AChE-biotests (Fig. 2) after the action of the ezerine and prozerine.

Table 1   Results of photometric analysis of AChE-biotests (see Fig.3)

Experiment		D *	A (%)
Normal incubation (D_n)		0.142±0,004	100
Incubation without substrate ($D_{w/s}$)		0.042±0.002	
Eserine	10^{-4}M	0.060±0.003	18
	10^{-5}M	0.095±0.005	53
	10^{-6}M	0.121±0.004	79
Proserine	10^{-4}M	0.059±0.002	17
	10^{-5}M	0.066±0.002	24
	10^{-6}M	0.134±0.002	92

*The number of the measurements is 9. Measurements of D was carried out at the TV-computer photometer (Suppl. 3)

Protocols of use: First register the image brightness of paper matrixes in plates (plate 1) then again register the second image of the plate with biotests (plate 2) (see suppl. 3). The value of D* was calculated by the standard formula:

$$D^* = \lg(S_{0j}/S_{ej}),$$

where, S_{0j} is the brightness of paper matrix (0-255 grey levels) in j-cell of plate 1, S_{ej} is the brightness of biotest in the same position on the plate (j*-cell of plate 2).

Statistical analysis of data: Analysis of matrix images was done by the program ScionImage (Scion Corp.,USA). The intensities of transmission light across matrixes were registered by means of a circle probe, its diameter being little less than matrix image. The program ScionImage, calculated the area of probe in pixel, the average brightness in the probe, the minimal and the maximal values of pixel brightness in the probe. The statistical analysis of results was carried out by program Sigma Plot 8.0.

Disadvantages: Only colourless extracts can be used for the reaction.

3.2   Formation of Yellow Product in Ellman Reaction

Principle: The determination of acetylthiocholine hydrolysis, based on the absorbance at 412 nm of the yellow product – complex of thiocholine with Ellman reagent.

Materials required

AChE preparations: Commercial acetylcholinesterase, horse (blood) serum preparation, human (blood) serum preparation, vegetative microspores of horsetail (*Equisetum arvense*) collected from meadows in April and May, generative microspores (pollen, male cells) or anthers of knights' star (*Hippeastrum hybridum*) grown in a green-house, the anthers from flowers of plant species belonging to the Liliales were chosen. *Lilium tenuifoliwn* Pisch. (Liliaceae) with red flowers and *Gladiolus hybridus* with white flowers (Iridaceae) were grown outdoors in the field. Knight's star (*Hippeastrum* sp.) and kafir-lily (*Clivia* sp.) hybrids (both from the Amaryllidaceae family) grown indoors, or cultivated cellular root tissue of ruta (*Ruta graveolens*). The anthers taken from the lily and gladiolus flowers were kept in a hermetically sealed vial at -24°C for 6 mon. Anthers from knight's star and kafir-lily were used immediately or kept in a hermetically sealed vial at 4°C for two wk; this did not influence the ChE activity.

Reagents: 20 mM Na-K-phosphate buffer pH 7.4, Ellman reagent 5,5"-dithio-bis(p-nitrobenzoic acid) shortly DTNB, or its red analogue 2,2-dithio-bis-(p-phenyleneazo)-bis-(1-oxy-8-chlorine-3,6) -disulfur acid in form of sodium salt, acetylthiocholine iodide or bromide, thiocholine.

Procedure: Cholinesterase activity was measured according to the modified biochemical methods developed for crude preparations (Gorun*et al.*, 1978), using Ellman reagent 5,5"-dithio-bis(p-nitrobenzoic acid) or its red analogue 2,2-dithio-bis-(p-phenyleneazo)-bis-(1-oxy-8-chlorine-3,6) disulfur acid in the form of sodium salt, which interact with thiocholine salt (Roshchina 2001). Water extracts of vegetative microspores of horsetail (*Equisetum arvense*) or *Hippeastrum hybridum* microspores (150 mg of microspores in 30 ml for 1 h) were used.

If anthers were chosen for the analysis, just before the experiments, the anthers were homogenized in 20 mM Na-K-phosphate buffer pH 7.4 with an electric homogenizer using a tenon-glass system at 0°C for 60 sec and then filtered through a nylon gauze. The enzyme - substrate reaction proceeded for 60 min at 25° C. Inhibitors was added to the mixture 10-20 min or 1 h before the substrate. Optical density of the samples at 412 or 620 nm was measured by Perkin -Elmer spectrophotometer. The incubation mixture contained 20 ml of homogenate, 10 ml of substrate solution and 170 µl of 20 mM Na-K-phosphate buffer (pH 7.4).The final volume 0.5 ml of the sample contain 20 mM Na-K-phosphate buffer (pH 7.4), 1 mM acetylthiocholine as a substrate and extract with the enzyme 0.2 ml. In the experiments with inhibitors, instead of 10 ml of buffer, 10 ml of inhibitor solution was added to the mixture 10 - 20 min before the substrate was added. The enzyme-substrate reaction proceeded for 60 min at 25 - 27°C. Simultaneously, mixtures containing 20 µl of homogenates and 180 µl of buffer were used as the reference solutions, when measuring the optical density of the samples. To measure a spontaneous hydrolysis of substrates, mixture containing 10 µl of substrate and 190 µl of phosphate buffer were used. The reaction was stopped by adding 1.8 µl of solution containing: 12.4 mg of Ellman reagent DTNB, 120 ml 96% ethanol, 50 ml 0.1 M Na-K-phosphate buffer (pH 7.4) and 50 ml distilled water. There are 2 control samples: either with enzyme or without enzyme. The optical density was measured with a spectrophotometer at 412 nm, using extinction coefficient EmM = 13.6. The activity of ChE was expressed in terms of mmoles of substrate hydrolyzed by 1 g of fresh mass (fr wt) or kg^{-1} of fresh mass s^{-1} ($s^{-1} kg^{-1}$ of fresh mass). The rates of hydrolysis were plotted against substrate concentrations (pS dependence). Michaelis-Menten constants (Km) aod maximal rates of hydrolysis (Vmax) were calculated using Lineweaver-Burk graphs and the Cornish-Bowden statistical method .

The activity of allelochemicals inhibitors of cholinesterase was assayed as listed below :

The concentration of inhibitor, causing 50% inhibition of enzyme activity (I_{50}, M) was calculated. In many cases the enzyme-inhibitor rate interaction constant (k_2 M^{-1} min^{-1}) was calculated according to the formula:

$$k_2 = (2,3/t[I]) \cdot \{\log (A_0/A_i)\}$$

where, A_i is the rate of the substrate hydrolysis in the presence of the inhibitor; A_0 is a rate of the substrate hydrolysis without inhibitor; (I) - is the inhibitor concentration, M and t is the time (min) of pre-treatment with the inhibitor before the substrate was added.

Observations: The reaction lasts 1 h. The results are expressed in μM $s^{-1}kg^{-1}$ of fresh mass. As shown in our experiments, the rates of hydrolysis of acetylthiocholine of studied plant cells varied from 0.4 ± 0.05 μM $s^{-1}kg^{-1}$ of fresh mass for pollen of knights'star up to 0.105 ± 0.01 μM $s^{-1}kg^{-1}$ of fresh mass for horsetail vegetative microspores. Table 1 shows I_{50} for inhibition of cholinesterase from microspores by used inhibitors.

Statistical analysis: Results of experiments are represented as $M\pm m$, where M – means, m – standard deviation of means: $m = \sigma/\sqrt{n}$, Where, σ is the standard deviation and n – is the population size.

Precautions: The yellow colour of the crude preparation may interfere with the Ellman reaction product, that need the appropriate special control.

Disadvantages: Dark coloured materials, due to the presence of anthocyanins, in particular are not suitable for the assay without strong dilution of the crude preparation

4. ANALYSIS OF ALLELOCHEMICALS

Principle: Cholinesterase is a cellular target for many toxins, including allelochemicals, pesticides and pharmaceuticals The effect of compounds on the enzyme activity is determined as the possible mechanism of the action on the cell.

Materials required: The cholinesterase preparation (biotests or simple extracts from biological tissues), the reagents for a reaction medium choosen (see section 3.2.), alkaloid physostigmine salycilate and its derivative neostigmine ("Sigma").

Procedure: Allelochemical and a compound belonging to natural artificial pesticides and medicinal drugs is preliminary added into the reaction media (see section Add). The difference in cholinesterase activity (measured as shown in sections 15.3) between a control (without the substance added) and the experimental variant is estimated. The results are compared with the effects of the cholinesterase inhibitors neostigmine and physostigmine.

Observations: The preliminary treatment of the cholinesterase-containing material with allelochemical (or other compound, e.g. active oxygen species, ozone free radicals and peroxides, formed in allelopathic relations) is for 30 min, then a substrate acetylcholinesterase is added to the reaction medium and final reaction of hydrolysis is for 1 h.

Experiment 1. *The effects of alkaloids and ozone on the rate of acetylcholine hydrolysis*
The cholinesterase activity is determined in the presence of alllelochemicals-alkaloids (Fig. 4) and reactive oxygen species ozone and peroxides (Table 1).

In the experiments with water extract from horsetail microspores (Table 1), all used alkaloids inhibited the acetylthiocholine hydrolysis. However, unlike acetylcholinesterase of eel, maximal inhibition was observed at concentrations 10^{-7}-10^{-5} M for variants with glaucine (70-75 %), lesser with sanguinarine (30-50 %), whereas with berberine, physostigmine and neostigmine was not higher 20-30 %. Since about 20-30 % rate of acethylthiocholine hydrolysis was inhibited by the animal cholinesterase inhibitors such as physostigmine and neostigmine (Augustinsson, 1963), therefore, the activity of microspores with pure cholinesterase was also approximately 20-30 %. In drastic inhibition, by glaucine and sanguinarine, the activity of non-specific esterase was also depressed. Main mechanism of

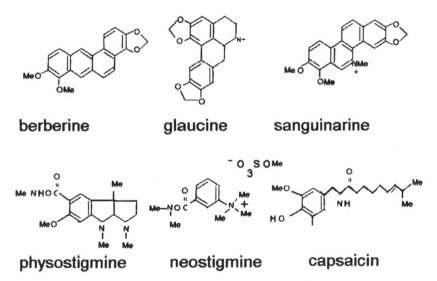

berberine   glaucine   sanguinarine

physostigmine   neostigmine   capsaicin

Fig. 4 Formulae of alkaloids wh Table 1. The rate (V) of the acetylthiocholine hydrolysis by water extracts (1:10 weight/volume) *Equisetum arvense* in the presense of allelochemicals –alkaloids and ozone

Table 1

Compound	$V \times 10^{-3}\ M\,kg^{-1}\ fr.ws^{-1}$	% of control
Controls without alkaloids	1.2 ±0.03	100± 4.2
Neostigmine 10^{-6}M	0.99±0.03	83±4.2
10^{-5}M	0.97±0.04	81±4.0
Physostigmine10^{-6}M	0.95±0.05	79±0.4
10^{-5}M	0.92±0.05	78±0.4
Berberine 10^{-6}M	0.99 ±0.04	82.6± 5
10^{-5}M	0.97±0.04	81±3.9
10^{-4}M	1.03±0.08	86±7.2
Glaucine 10^{-7}M	0.30±0.03	25±0.3
10^{-6}M	0.40±0.009	33±0.1
10^{-5}M	0.40±0.0009	33±0.1
10^{-4}M	0.38±0.05	30±5.0
Sanguinarine 10^{-7}M	0.85±0.03	70±3.0
10^{-6}M	0.77±0.04	64±4.0
10^{-5}M	0.53±0.04	44±4.0
10^{-4}M	0.83±0.06	69.6±6.5
Ozone 0.8×10^{-9}M	1.2±0.06	100±6.5
2×10^{-9}M	0.67±0.06	57.5±6.5
4×10^{-9}M	0.97±0.06	78±6.5
8×10^{-9}M	0.99±0.06	80±1

the cholinesterase inhibition may be through binding with anionic centre of the enzyme (quarternary nitrogen atom), as shown for protoberberine and benzophenantridine alkaloids (Ulrichova *et al.*, 1983). Ozone and peroxides as active oxygen species are formed in living cells at normal and at allelopathic relations (Roshchina and Roshchina, 2003) decreased the rate of acetylthiocholine hydrolysis by 20-30 %.

The sensitivity of the cholinesterase of microspores to allelochemicals analyzed, may be also estimated from the comparison of I_{50} (a concentration of the inhibitor, which decreases the rate of the enzyme synthesis by 50 % of the control). As seen from Table 2, among the studied compounds, the highest sensitivity to neostigmine was observed. Rutin-like phenolic compound, flacosid from *Phellodendron amurense*, also inhibits the cholinesterase activity, at higher concentrations. Organic peroxide *tert*-butylhydroperoxide and alkaloid physostigmine were equally effective in similar concentrations. To compare the cell sensitivity, pollen grains of *H.hybridum* were most sensitive than vegetative microspores of *E.arvense*, because the effect of inhibitor is observed at lower concentrations.

Table 2 I_{50} for inhibition of cholinesterase from microspores by biologically active compounds used.

Inhibitor	I_{50}, M
Vegetative microspores Equisetum arvense	
+ Neostigmine	0.9×10^{-5}
+ Flacosid	6×10^{-4}
+ *tert*-butylhydroperoxide 10^{-4}M	1.1×10^{-4}
Pollen of *Hippeastrum hybridum*	
+ Neostigmine	5×10^{-6}
+ Physostigmine	10^{-4}

* Flacosid = 7-β- D-glucopyranoside-8(3-methylbut-2-enyl)-4',5,7 –trioxyflavonone from *Phellodendron amurense* Rupr. (family Rutaceae)

5. SUPPLEMENTS

5.1   Methods of Photometric Analysis of AChE-Biotests Image

5.1.1   Experimental device for photometric measurements

The experimental device for photometric measurements is shown schematically in Fig. 7. The device includes a measuring box (1) in which a mobile platform (2) is situated with two windows (3), into which plates (3 ') are inserted, each plate having 9 holes with paper matrixes.

The plates are illuminated with an illuminator (4) consisting of a halogen lamp (15 V, 150 W), a lens system and an interference filter (l = 483nm) and a power unit (5). The image of plates with matrixes is registered with a video camera WAT-505EX (6) (Watec, Japan; 795ö596 px, 0.01 lux) supplied with an objective Y1304M (Jamano, Japan, 1/3", 4 mm, F1,2, 65°) and a power unit (7). The image is captured on the computer (Pentium 3, 64 RAM, 16 Mb) (8) by a FlyVideo (AnimationTechnologies Inc., USA, ttp: // www.lifeview. com.tw) as an 8-bit image (256 intensity levels of grey scale). There is an additional device for direct photometric measurements of light intensity (9) (photoresistor, CdS, 10 mm diameter, connected conventionally for photometric measurements.

5.1.2   Protocols of use

For the calculation of biotests optical density and computer analysis see 1.3.1 Experiment No. 3.

5.2   Direct Measurement of Optical Parameters of Biotests

Photometric measurements of colouring in the AChE-biotests after biochemical hydrolysis of acetylthiocholine may be made by using any spectrophotometer or special microplate photometers. Spectrophotometry allows us to measure the reflection, coefficient of the AchE-biotests or optical density (see Chapter 11, 3.1, Exp. 2). As a blank for the measurement of red product (the absorbance maximum 480 - 490 nm) in the Karnovski-Root reaction, it is possible to use Congo red dye with a maximum of absorption about 500 nm (see Chapter 11, 3.1, Exp. 2).

5.2.1   Microplate photometers

The use of the 96-cells plate and microplate photometers represents a convenient and fast way of quantitative photometric analysis of reactions of the chemical tests and biotests made on the basis of paper materials. The firm microplate photometers are supplied with the necessary software and systems of scanning a plate which carry out not only one-wave, but also the multiwave photometric analysis, that will enable us to increase the accuracy of the analysis.

The measurement of optical density in AChE-biotests after analytical procedure can be done on special microplate photometer, for example, MicroReader 4 (*Hyperion Inc.*, USA). The measurement of optical density was at $\lambda = 490$ nm. For installation of microplate photometer parameters, the software of the device is used. According to our data, the best results were found with the use of modified 96-cells plate (Budantsev and Budantseva, 2005). They differ from a standard plastic plate by absence of cells bottom.

5.2.2   Indirect smeasurement of optical parameters AChE-biotests (TV-computer analysis)

TV-detection of paper matrix optical density has a number of advantages as compared with direct photometric measurements using photoelectronic devices (a photomultiplier photometer). In the former case, a scientist registers the image of the matrix and computer analysis enables us to receive not only the integral estimation of light transmission, but also the scatter in the intensity of light transmitted through the matrix. In the latter case, we obtain only the integral characteristic of transmitted light. Another advantage of TV-detection consists in simultaneous measurements of large number of matrixes and a simple input of matrix images to the computer. Programs of ScionImage type allow necessary photometric parameters to be quickly and effectively calculated. Finally, this method makes possible simple estimation of the 'optical homogeneity' of paper materials. This parameter is of importance in choosing and designing materials for paper matrixes in the development of sensitive paper-tests.

The description of special device for TV-computer photometry is given in Suppl.3.

5.3   Flow Column Reactor

5.3.1   General design of reactor

The design of one reactor module is shown in Fig. 5. The water jacket (the thermostat) of colomn reactor is included in two plates (1), outside glass cylinder (2) and four studs (3). The silicone gasket is between the plates and glass cylinder .

In the top and bottom plates there are unions (4) for connection with ultratermostst. The internal tube of thermostat (5) with the help of silicone gaskets and lock-nuts is fixed on the top and bottom plates of the module.

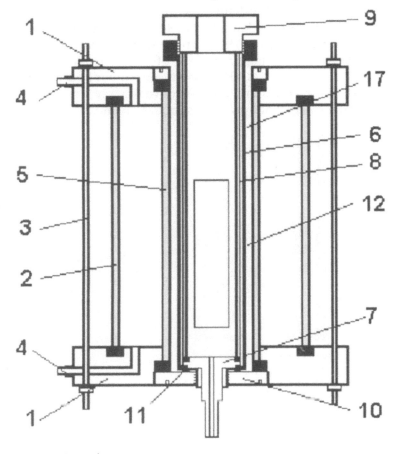

Fig. 5   The design of one reactor module (details in text)

The column reactor is located in the internal cylindrical space of the thermostat. The external case of reactor (6), made from stainless steel has teflon shell in the bottom part (7), on which is pressed through silicone gasket the glass cylinder (8), representing an internal volume of a reactor. In walls of the external case, there are windows for the visual control of internal reactor space. On the top part of the reactor case, there is a lock-nut (9), pressing the internal cylinder of reactor to a silicone gasket on the teflon shell (7). Thus internal reactor space is reliably separated from the external case (6). The external reactor case is connected with the bottom plate by threaded connection. Between the external reactor case and nut (10) there is silicone gasket (11) for isolation of internal space (12) between the thermostst and reactor case. Thus in the internal space of the thermostat it is possible to replace easily and quickly the flowing reactor (detail 6-9, 11) through threaded connection of the external case with a nut (10).

In Fig. 6 the scheme of packing of the biotests in internal reactor volume is shown. The kapron mesh (km) is put on the surface of the Teflon shell (7) . Then is stacked silicone ring (sr), biotest (bt), then the following ring and biotest etc. The diameter of AChE-biotests is equal to our installation 7.6 mm, thickness of silicone rings 1 mm, in reactor we stacked a package from 10 biotest, common height of a package is equal to 10 mm. After packing the

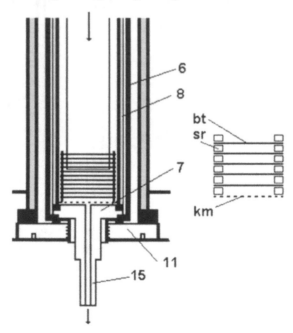

Fig. 6   Scheme of packing of the biotests in reactor (details in text)

biotests the reactor is connected with nut (10). The column reactor is ready for experiments.

In Fig. 6, the scheme of formation from modules of the multichannel block is shown. The few thermostats are fixed on a stand (13) and the thermostats unions consistently incorporate among themselves and with ultrathermostat (14). The ends (15) of teflon shells (7) are connected with a multichannel peristaltic pump (16).

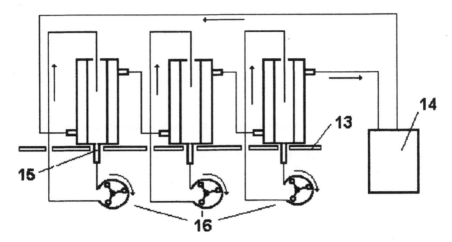

Fig. 6   Scheme of formation of the multichannel flow system

Our device includes the 5 column reactors, 5-channels peristaltid pump (pp1-05, Poland) and ultra-thermostat U-1 (Germany). In the work space between the internal tube of thermostat (5) and external reactor case (6) water is filled by a backlash (17) in the top part of the module (fig.1).

The described module is completely folding, which allows it to quickly replace its parts and to clean all elements of the module, that is important in work with poisonous and radioactive substances.

5.3.2   The protocol of the analysis

10 AChE-biotests are located in each column reactor. Through a hole in the lock-nut (9) of a column 3 ml of icubation solutions are injected (see chapter 15, Section 3.1). The speed of flow is 1.4 ml/min in the closed flowing system. Time of incubation is 2 h at room temperature. After incubation, the columus is disconnected, AChE-biotests are taken, dried and their photometric analysis is done.

The protocol of experiments includes the following: a) the incubation in complete incubation solution («standart incubation», 1-st channel of flow

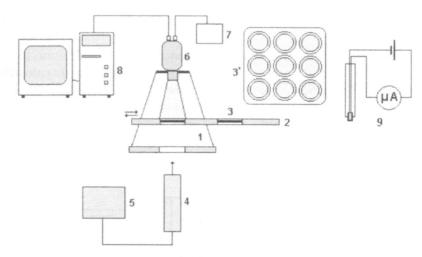

Fig. 7 Scheme of a photometric analysis of the biotest images (Budantsev, 2004).

system); b) the incubation without substrate («control of reaction specificity», 2-nd channel of flow system); c) the incubation with different concentration of inhibitors («experimental probes», 3-5th channels of flow system) (see 1.3.1 Experiment 3).

6. REFERENCES

Augustinsson, K.B. (1963). Classification and comparative enzymology of the cholinesterases and methods for their determination. In : *Cholineasterase and Anticholinesterase Agents* , G.B. Koelle (ed). *Handbuch der Experimentale Pharmacologie.* Springer-Verlag, Berlin, Germany**15**: 89-128.

Atta-ur-Rahman, Parveen, S., Khalid, A., Farooq, A., and Choudhary, M.I. (2001). Acetyl and butyryl cholinesterase -inhibiting triterpenoid alkaloids from Buxus papillosa. *Phytochemistry* **58** : 963-968.

Barbour, J.D., Farrar Jr., R.R. and Kennedy, G.G. (1993). Interaction of *Manduca sexta* resistance in tomato with insect predators of *Helicoverpa zea. Entomologia Experimentalis Et Appilcata* **68**: 143-155.

Bell, H. A., Down, R.E., Fitches, E.C., Edwards, J.P. and Gatehouse, A.M.R. (2003). Impact of genetically modified potato expressing plant derived insect resistance genes on the predatory bug, *Podisus maculiventris* (Heteroptera : Pentatomidae). *Biocontrol Science and Technology* **13**: 729-741.

Botre, F., Lotenti, G., Mazzei, F., Simonetti, G., Porcelli, F., Botre C. and Scibona, G. (1994). Cholinesterase based bioreactor for determination of pesticides. *Sensors and Actuators.* Ser.B. **18-19**: 689-693.

Budantsev, A.Yu. (2004). Photometric determination of compounds in paper matrices using digital imaging technique and transmitted light images. *Journal of Analytical Chemistry* (Russian) **59**: 791-795.

Budantsev, A. Yu. (2005). Signal biosensors as identifiers. *Advantages in Modern Biology* (Russia) **125**: 67-78.

Budantsev, A. Yu. and Budantseva, T.A. (2005). Photometric analysis of paper tests using microplate readers. *Journal of Analytical Chemistry* (Russsian) **60**: 794-797.

Budantsev, A. Yu., and Litvinova, E.G. (1996). Test-paper for primary screening of pharmacological compounds possessing MAO-inhibitory activity. *Experimental and Clinical Pharmacology* (Russia) **59**:68-70.

Budantsev, A. Yu., Litvinova, E.G. and Kovaleva, M.A. (1997). Indicator paper for biogenic amines. *Journal of Analytical Chemistry* (Russia) **52**: 539-542.

Budantsev, A. Yu., Litvinova, E.G. and Kovaleva, M.A. (1998). Biosensors on the basis of tissue enzyme preparations. *Sensors Systems* (Russian) **12**: 42-55.

Budantsev, A. Yu. and Roshchina, V.V. (2005). Testing of alkaloids inhibitory activity to acetylcholinesterase. *Plant Resourses* **41**: 131-138. (In Russian).

Evtugyn, G.A., Budnikov, H.C. and Nikolskaya, E.B. (1999). Biosensors for the determination of environmental inhibitors of enzymes. *Advantages of Chemistry* **68**:1142-1167. (In Russian)

Gorun, V., Proinov, I., Baltescu, V., Balaban, G. and Barzu, O. (1978). Modified Ellman procedure for assay of cholinesterases in crude enzymatic preparations. *Analytical Biochemistry* **86** : 324-326.

Karnovsky, M. and Roots, L.A. (1964). A direct colouring method for cholinesterases. *Journal of Histochemistry and Cytochemistry* **12** : 219-225.

Narahashi, T. (Ed) (1979). *Neurotoxicology of Insecticides and Pheromones.* Proceedings of Symposium on *Chemistry of Neurohormones and Neurotransmission.* American Chemical Society 175th Anniversary, March 14-15, 1978. Anaheim, California. Plenum Press, New York, USA.

Roshchina, V.V. (2001). *Neurotransmitters in Plant Life.* Science Publishers: Einfield, New Hampshire, USA. .

Roshchina V.V. (2004). Plant cholinesterase activity as a biosensor: cellular models. In: *Cholinergic Mechanisms.* I. Silman, H. Soreq, L. Anglister, D.M. Michaelson and A. Fisher (eds). pp. 260-263. Martin Dunitz. London, UK.

Sternfeld, M., Ming, G., Song, H., Sela, K., Poo, M. and Soreq, H. (1998). Acetylcholinesterase enhances neutrite growth and synapse development through alternative contributions of its hydrolytic capacity, core protein, and variable C termini. *Journal of Neuroscience* **18** : 1240-1249.

Ulrichova, J., Walterova, D., Preininger, V., Slavik, J., Lenfeld, J., Cushman, M. and Simanek, V. (1983). Inhibition of acetylcholinesterase activity by some isoquinoline alkaloids. *Planta Medica* **48** :111-115.

Wenk H.von, Krug H., und Fletcher A.M. (1973). Microspectrometric method for quantitative determination of the acetylcholineaterase activity). *Acta Histochemistry* **45**: 37-60. (In German).

Chapter

Methods to Study Effect of Allelochemicals on Algae Membrane Integrity

F.M. LI[1,2] and H.Y. Hu[2]

1. INTRODUCTION

Elucidation of the mode of allelochemicals on algae membrane is an important but challenging task, in allelopathic research for algae control, because of the multiple potential molecular targets. Different kinds of allelochemicals may act on different target sites of algae membrane. Some allelochemicals may cause rapid plasma membrane leakage in acceptors i.e. changes in the cell membrane may occur. The composition of cell membrane lipid fatty acids is determined to study the allelochemicals mode of action mode. The allelochemicals may cause oxidation of cell membrane lipid fatty acids.

Oxidation of lipid fatty acids may be caused by the excessive reactive oxygen species in the plasma. In photosynthetic organisms, environmental stress can create oxidative stress through increased production of reactive oxygen species (ROS), e.g. singlet oxygen ($^1O_2$), the hydroxyl radical (HO·) and hydrogen peroxide (H_2O_2). Reactive oxygen species may cause lipid peroxidation, while superoxide dismutase (SOD), catalase(CAT) and peroxidase (POD) inhibit lipid peroxidation. Under oxidative conditions, microalgae respond by increasing antioxidant defences, notably enzymes

[1]College of Environmental Science and Engineering, Ocean University of China, Qingdao 266003, China. E-mail: lfm01@mails.tsinghua.edu.cn.
[2]Environmental Simulation and Pollution Control State Key Joint Laboratory, Department of Environmental Science and Engineering, Tsinghua University, Beijing 100084, China. E-mail: lfm01@mails.tsinghua.edu.cn

such as SOD, POD and CAT. In most cases, these defence mechanisms work jointly to reduce the levels of ROS in the cells. Generally, stress-tolerant species have a more effective defence system against ROS than stress-susceptible species. The activities of antioxidation enzymes affected by adding allelochemicals should be determined to elucidate the mode of action of antialgal allelochemicals. A sharp decrease in antioxidation enzyme activity due to addition of allelochemicals suggest an increase of ROS.

2. EFFECTS OF ALLELOCHEMICALS ON ALGAL ANTIOXIDANT ENZYMES

Pre-culture the algae to about 10^8cells/ml. Then, add the allelochemicals to the culture, followed by incubation for 24 h. Then measure the activities of SOD, POD of algae with the addition of different concentrations of allelochemical. A control (without allelochemical) should run in parallel.

Experiment 1. *Evaluation of the effects of allelochemicals on the activity of algal antioxidant enzymes.*
The algae are pre-cultured to about 10^8cells/ml. Then the allelochemical is added to the culture, followed by incubation for 24 h. Effects of different concentrations of allelochemicals on the activities of SOD, CAT, POD of algae are studied. A control (without allelochemical) should run in parallel.

2.1 Effects of Allelochemicals on Algal Superoxide Dismutase (SOD) Activity

Materials required
Buffer and chemicals: pH 7.8 phosphate buffer (16.29g $Na_2HPO_4\cdot2H_2O$, 1.17g $NaH_2PO_4\cdot H_2O$, 1000 ml distilled water), Ethylene Diamine Tetra Acetic Acid (EDTA), Methionine (Met), Polyvinylpyrrolidone (PVP), Triton X-100, Phenylmethylsulphonylfluoride (PMSF), Riboflavin, Nitro Blue Tetrazolium (NBT).

Apparatus: Centrifuge, spectrometer.

Procedure: Extracts for measuring SOD activity (EC 1.15.1.1) are made by grinding algal cells in 4 ml of 0.1M phosphate buffer(pH 7.8), containing 0.1 mM EDTA, 1% PVP, 1% Triton X-100, 1 mM PMSF. Centrifuge the crude homogenate at 10,000g for 15 min and immediately assay the SOD activity. Monitor the inhibition of the photochemical reduction of nitro blue tetrazolium (NBT) to measure the SOD activity as per method of Paoletti. In the 3.0 ml reaction mixture [containing 0.013mol/L Met, 1.3×10^{-6}mol/L riboflavin, 7.5×10^{-5} mol/L NBT, and 0.05 mol/L phosphate buffer (pH 7.8)] add 50 µl of crude enzyme extract. Initiate and last the reaction for 15 min (37°C) under 4000 lx fluorescence light. Record the optical density at 560 nm. In control, replace the crude enzyme extract by phosphate buffer. One unit of SOD is defined as the increase in A_{560} per minute per 10^8 algal cells.

2.2 Effects of Allelochemicals on Algal Peroxidase (POD) Activity

Materials required

Chemicals: Ascorbate, Potassium Phosphate, Ethylene Diamine Tetraacetic Acid (EDTA), Sodium Ascorbate, H_2O_2, Polyvinylpyrrolidone (PVP).

Apparatus: Spectrometer.

Procedure: The Ascorbate peroxidase (EC 1.11.1.7) activity can be obtained by measuring the oxidation of ascorbate in the presence of H_2O_2. Grind the algal sample in liquid nitrogen and extract in 2.5 ml 50 mM potassium phosphate buffer (pH 7.0) containing 10% (w/v) polyvinylpyrrolidone (PVP), 0.25% Triton X-100 and 0.5 mM ascorbate. Then centrifuge the extraction for 5 min at 14,000 rpm at 4°C to eliminate debris (Sigma Laborzentrifugen). Initiate the oxidation of ascorbate by adding 50 µl supernatant to 950µl potassium phosphate buffer (pH 7.0, 50 mM) [containing 0.1 mM EDTA, 0.5 mM ascorbate and 0.1 mM H_2O_2]. After 30s of the reaction, measure the decrease in absorbance at 290 nm with a spectrophotometer. Perform the assay at 20°C. Define one unit of POD as the variance of absorbance per 10^7 cells (** A_{290} / 10^7 cells).

3. EFFECTS OF ALLELOCHEMICALS ON ALGAL PLASMA MEMBRANE LIPID FATTY ACIDS

Materials required

Chemicals: KOH, Petroleum Ether (60–80°C fraction), Methanol, HCl, Boron Trifluoride(BF_3).

Apparatus: Sonic apparatus (Soniprep 150, MSE), Gas Chromatography / Mass Spectrometry (GC/MS).

Procedure: Harvest the algal cells by centrifugation at 3000 rpm for 5 min at 4°C, then, wash the cells in distilled water thrice. Prepare three replicates for each treatment. Sonicate the algal cells under nitrogen condition at 14 MHz for 10 min to extract the membrane fatty acids. Add 2 ml of 2 M KOH in 95% (v/v) methanol and 2 ml of benzene to the sonicated cells. Fill the headspace with N_2 gas and incubate the mixture at 80°C for 3 h. On cooling, add an equal volume of methanol and extract the non-saponified fraction by shaking with three successive 5 ml aliquots of petroleum ether (60–80°C fraction). Acidify the lower methanol layer with 6 M HCl and extract the saponified lipids with petroleum ether (three aliquots of 5 ml). Evaporate the saponified lipid fraction, then derive it to methyl esters by the addition of 1 ml of 14% BF_3 in methanol in a cuvette. Fill the headspace of the cuvette with N_2 gas, seal and heat it at 80°C for 1 h. Add the contents to 5 ml distilled water and extract three times with 5 ml petroleum ether. Concentrate the pooled ether fractions. Analyze the fatty acids in the ether fraction with GC-MS.

4. EXAMPLES

In our previous research, we found that the antialgal allelochemical Ethyl 2-Methylacetoacetate (EMA) caused loss of cell membrane integrity. It hinted that EMA may cause a change in the membrane. It is reported that environmental stress may increase the concentration of ROS in plant cell. The excessive ROS may cause a decrease of antioxidation enzymes activity and lipid peroxidation. The effect of EMA on the activity of SOD and POD and lipid fatty acids of *Chlorella pyrenoidosa, Chlorella vulagaris* and *Microcystis aeruginosa* were evaluated to elucidate the mode of action of EMA.

4.1   Effects of Ethyl 2-Methylacetoacetate (EMA) on the Algal Antioxidant Enzymes Activity

The activities of antioxidant enzymes of *Chlorella pyrenoidosa, Chlorella vulagaris* and *Microcystis aeruginosa* with the addition of EMA were investigated. A low concentration of EMA (0.25mg/L) caused increases in SOD and POD activities (Fig. 1). Higher concentrations of EMA (above 0.5mg/L) decreased the SOD and POD activities of *C. pyrenoidosa* and *M. aeruginosa* but caused no apparent decrease in the SOD or POD activities of *C. vulagaris*. These results indicated that the added allelochemical, which act as an environmental stress, created oxidative stress through increased production of reactive oxygen species (ROS). In oxidative conditions, the algae responded by increasing antioxidant defences, including superoxide dismutase (SOD) and peroxidase (POD). But higher concentrations of allelochemical and the excessive ROS it creates, decreased the SOD and POD activity. The excessive reactive oxygen species may cause lipid peroxidation.

4.2   Effects of EMA on Membrane Fatty Acids

Fatty acids were extracted following the above procedure. Fatty acids are analyzed by GC-MS. The GC used was a Perkin-Elmer 8600 fitted with flame ionization detector. An SE-54 quartz capillary column (30m) was used. The

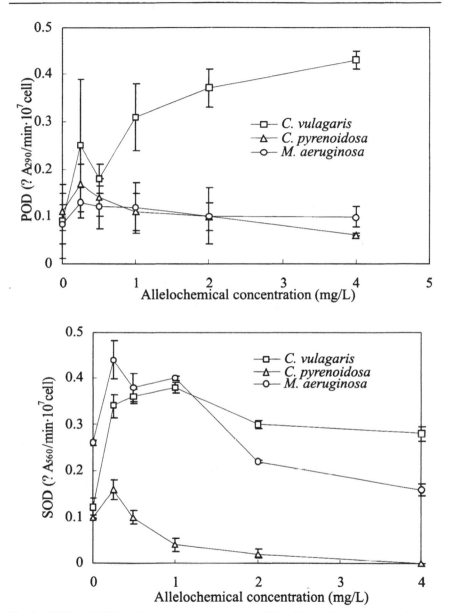

Fig. 1 POD and SOD activity changes caused by EMA. All error bars correspond to the standard deviation.

column temperature was initially held at 60°C for 2 min, and then programmed to 300 at a rate of 5 min. The injector temperature was maintained at 280°C, the injection volume was 1 ml with the splitless mode.

Table 1   The effect of EMA on fatty acids extracted from C. pyrenoidosa , C. vulagaris and M. aeruginosa

	Lipid	% of total fatty acids		
		C. pyrenoidosa	C. vulagaris	M. aeruginosa
with no allelochemical	C14:0 Myristic acid	16.57±0.35	8.17±0.12	0±0
	C16:2 Hexadecadienoic acid	0±0	0±0	11.93±0.44
	C16:0 Cetylic acid	33.68±0.09	26.95±0.21	16.8±0.22
	C18:3 Linolenic acid	23.31±0.64	29.56±0.29	0±0
	C18:2 Linolic acid	11.46±0.81	30.41±0.06	30.26±0.25
	C18:1 Oleinic acid	0±0	0±0	18.85±0.14
	C18:0 Stearic acid	14.98±0.32	4.9±0.18	22.15±0.31
with 2mg/L allelochemical	C14:0 Myristic acid	4.66±0.06	7.77±0.05	0±0
	C16:2 Hexadecadienoic acid	0±0	0±0	7.21±0.04
	C16:0 Cetylic acid	22.04±0.13	25.04±0.37	11.4±0.27
	C18:3 Linolenic acid	37.68±0.24	29.17±0.22	0±0
	C18:2 Linolic acid	25.91±0.51	32.1±1.02	42.88±1.54
	C18:1 Oleinic acid	0±0	0±0	28.46±0.33
	C18:0 Stearic acid	9.71±0.05	5.91±0.65	10.05±0.03

Data are presented as mean percentages of total fatty acids from triplicate experiments; S.D.= standard deviation.

Constituents were identified by peak matching against standards in the NIST 95 Computer Library. The relative amounts of constituents were calculated by integrating all peaks with areas greater than 1%.

The major components of membrane fatty acids of cells of *C. pyrenoidosa* and *C. vulagaris* were C14:0, C16:0, C18:0, C18:2, C18:3 and those of *M. aeruginosa* were C16:0, C16:2, C18:0, C18:1, C18:2. Table 1 shows the effect of EMAon membrane fatty acid composition, expressed as the mean percentage of each lipid (C14:0, C16:0, C16:2, C18:0, C18:1, C18:2, C18:3) extracted from cells after incubation with 2mg/L EMA. EMA caused the proportion of unsaturated fatty acids (C18:2, C18:3 in *C. pyrenoidosa* and *C. vulagaris*, C16:2, C18:1, C18:2 in *M. aeruginosa*) to increase, while saturated fatty acids decreased. The changes in the fatty acid composition indicated a change in membrane fluidity. All the changes were statistically significant (P<0.01).

5. SUGGESTED READINGS

Baranenko, V.V. (2001). Pea chloroplasts under clino-rotation: lipid peroxidation and superoxide dismutase activity. *Advances in Space Research* **27**: 973-976.

Bhowmik, P.C. and Inderjit (2003). Challenges and opportunities in implementing allelopathy for natural weed management. *Crop Protection* **22** : 661-671.

Dayan, F.E., Romagni, J.G. and Duke, S.O. (2000). Investigating the mode of action of natural phytotoxins. *Journal of Chemical Ecology* **26**: 2079-2094.

Li Fengmin, Hu Hongying (2005). Isolation and Characterization of a Novel Antialgal Allelochemical from *Phragmites communis*. *Applied and Environmental Microbiology* **71**: 6545-6553.

Mishra, N.P., Mishra, R.K. and Singhal, C.S. (1993). Changes in the activities of antioxidant enzymes during exposure of intact wheat leaves to strong visible light at different temperatures in the presence of protein synthesis inhibitors. *Plant Physiology* **102** : 903-910.

Rose, A.H. and Veazey, F.J. (1988). Membranes and lipids of yeasts. In: *Yeast: A Practical Approach*, .,I. Campbell and JH Duffus (eds). pp. 255–275. IRL Press Oxford: Washington, DC, USA

Chapter

Total Phenolics and Phenolic Acids in Plants and Soils

L. Djurdjević *, M. Mitrović and P. Pavlović

1. INTRODUCTION

During allelopathic investigations of interrelationships among different plant species in their natural sites, special attention is paid to the presence of phenolic acids (PAs) in dominant plants, plant litter and in the soil. They are localized in all plant organs, both as free and in bound forms i.e. associated with other compounds such as lignin (Kögel and Bochter 1985; Kögel 1986) and polysaccharides of cell walls (Whitmore 1976). Phenols are transferred from plants to the litter and soil by foliage and stem leaching, leaf fall, root exudates, microbial decomposition of plants remains, (mainly the degradation of lignin). PAs have been identified as the main phytotoxins of some plant species (Lodhi and Rice 1971; Chou and Muller 1972). Phytotoxins play a significant role as germination inhibitors that affect growth of seedlings, soil nitrification and nitrifying bacteria (Lodhi and Killingbeck 1980, Einhellig *et al.* 1982, Khan and Ungar 1986). This chapter describes the methods for extraction, detection and measurement of total phenolics and phenolic acids in dry plant material (vegetative plant parts, mostly leaves), plant litter and soil (organic and mineral); method of induced chlorophyll fluorescence kinetics of photosystem II for determination of photosynthetic efficiency; methods for chlorophyll content determination and methods of biotests.

Department of Ecology, Institute for Biological Research "Siniša Stanković", University of Belgrade, Bulevar Despota Stefana 142, 11060 Belgrade, Republic of Serbia.* E-mail: kalac@ibiss.bg.ac.yu

2. EXTRACTION OF PLANTS PHENOLIC ACIDS WITH AMBERLITE IR-45 (OH) RESIN

Several organic solvents, acids or bases have previously been used for both qualitative and quantitative analyses of phenolic inhibitors, which were extracted for different periods of time (1-48 h) either at room temperature or using a Soxhlet apparatus (Ishikura 1976, Nagels and Parmentier 1976). However, it is believed that such methods do not provide correct data either on the amount of phenolic compounds or about their qualitative composition. According to Swain (1976) long-lasting extraction in the Soxhlet apparatus was unsuitable for the extraction of most flavonoids. The efficiency of PAs extraction from dry plant material with ethanol alone and with ethanol in the presence of Amberlite IR-45 (OH) ion-exchange resin was compared. The procedure with Amberlite IR-45 (OH) ion-exchange resin includes neither heating nor use of acids and bases.

Principle: Instead of long-lasting extraction with different organic solvents, acids or bases in the Soxhlet apparatus is unsuitable for the extraction of most flavonoids, the Amberlite IR-45 (OH) ion-exchange resin was used for the extraction of plant phenolics.

Materials required: Dry plant material, methanol, diethyl ether, anhydrous Na_2CO_3 and 2N HCl (HPLC grade), N_2, distilled water, Amberlite IR-45 (OH) ion-exchange resin (Rohm and Haas Company, Philadelphia, USA), graduated tube, Erlenmeyer flask (100 ml), filter paper, funnels, separatory funnels, rotary evaporator, mill, water bath, shaker, sieve of 0.5 mm in diameter, pipette of 1, 2 and 10 ml, test tube of 20 ml.

Procedure: Dry blackberry (*Rubus hirtus* W.K.) leaves were used as the source of PAs. Plant material was pulverized and sieved through a 0.5 mm pore size screen. Two procedures for extraction were applied: (i) 1.0 g of pulverized plant material was extracted with 30 ml of 80% (v/v) methanol HPLC grade and (ii) 1.0 g of pulverized plant material was mixed with 30 ml of 80% (v/v) methanol and 3.0 g of Amberlite IR-45 (OH) ion-exchange resin (Rohm and Haas Company, Philadelphia, USA). Extraction was done in 100 ml Erlenmeyer flask with constant shaking (24 h) and three exchanges of methanol. Methanol extracts were pooled, vacuum evaporated under N_2 and the dry residues were hydrolyzed in 2N HCl for 60 min in a boiling water bath. Hydrolyzates were extracted with diethyl ether. The ethereal phase was dehydrated with anhydrous Na_2CO_3 and evaporated to dryness in a stream of nitrogen. The dry residues dissolved in 80% methanol were further subjected to spectrophotometry (Shimadzu UV 160 spectrophotometer) for total phenolics and high-performance liquid chromatography

(HPLC) for phenolic acids analysis. When procedure (ii) was applied, the ion-exchange resin was separated from the methanol phase and eluted with three 40 ml aliquots of 80% methanol. The resin bead eluates were evaporated to dryness and subjected to spectrophotometry (Shimadzu UV 160 spectrophotometer) for total phenolics and high-performance liquid chromatography (HPLC) for phenolic acids analysis.

Experiment I. *Determination of total phenolics spectrophotometrically*

Materials required: Sample solution in methanol, distilled water, 20% Na_2CO_3, Folin-Ciocalteu's phenol reagent, water bath, graduated tube, cuvette, micro pipette (0.1 ml), pipette of 1, 2 and 10 ml, test tubes of 20 ml, pure ferulic acid (Serva, Germany), Shimadzu UV 160 spectrophotometer.

Procedure: A 0.002 ml aliquot of the sample solution in methanol was taken and 7 ml distilled water plus 0.1 ml Folin-Ciocalteu's phenol reagent was added and after 3 min 0.2 ml of 20% Na_2CO_3 was included. After boiling at $90\,^0C$ (exactly 5 min) samples were cooled at room temperature and were diluted with H_2O to 10 ml volume. Only distilled water and reagents were used as a blank. The absorbance of total phenolics was measured at 660 nm spectrophotometrically (a Shimadzu UV 160 spectrophotometer) as per Feldman and Hanks (1968), with a sensitivity of $0.05\,\mu g/g$ d.w. A standard curve was constructed with different concentrations of ferulic acid (Serva, Germany). Concentrations of ferulic acid varied from $0.33\text{-}80\,\mu g/ml$ (Table 1).

Table 1 Total phenolics and phenolic acids content ($\mu g/g$) in dry *Rubus hirtus* leaves determined by two different methods

Sample	Phenolic acids					
	p-Coumaric	Ferulic	p-Hydroxy benzoic	Vanillic	Total phenolic acids	Total phenolics
A-Control	164±14	117±19	29±3	131±24	441±43	14112±1500
B-Unbound to Amberlite	101±9	94±19	26±4	82±12	303±21	10384±1121
C-Eluted from Amberlite	169±11	82±9	25±4	61±16	336±46	12987±1384
B+C	269±17	176±17	51±4	143±10	639±46	23371±2678
% of control	164	151	173	109	145	166

A-Phenolic acid's extraction with methanol alone.

B, C-Phenolic acid's extraction with methanol in the presence of Amberlite IR-45 (OH) ion-exchange resin

Statistical analysis: The data were analyzed using t-test for dependent variables or when have large sample or more than two combinations one-way ANOVA.

Experiment II. *Determination of phenolic acids by HPLC analysis*

Materials required: Acetonitrile, o-phosphoric acid (HPLC grade), Nucleosil 100-5 C_{18} column (5 µm; 4.0x250 mm-Agilent Technologies, USA), vial, graduated tube, cuvette, distilled water, pure p-hydroxybenzoic and syringic acid (Acros Organics, U.S.A.), ferulic, vanillic and p-coumaric acid (Serva, Germany), HPLC (Hewlett Packard HP 1100).

Procedure: Phenolic acids were detected between 210 and 360 nm using a Hewlett Packard diode array detector (HP 1100 HPLC system). The separation was achieved with a Nucleosil 100-5 C_{18} column; 5 mm; 4.0x250 mm (Agilent Technologies, USA) at a flow rate of 1.0 ml/min and injection volume of 5 µL. For the elution, a discontinuous acetonitrile-water gradient was used: 15% acetonitrile (5 min), 30% acetonitrile (20 min), 40% acetonitrile (25 min), 60% acetonitrile (30 min), 60% acetonitrile (35 min) and 100% acetonitrile (45 min, isocratic). To avoid tailing of the phenolic acids, 0.05% o-phosphoric acid was added to the solvents. Phenolic acids were identified on the basis of retention time values and absorption peaks of pure p-hydroxybenzoic and syringic acid (Acros Organics, U.S.A.), ferulic, vanillic and p-coumaric acid (Serva, Germany), which served as references.

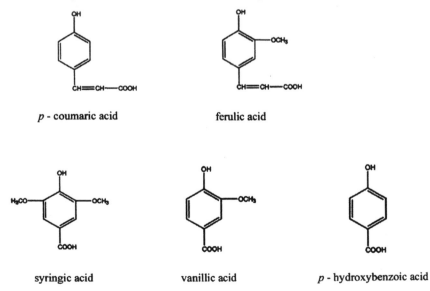

p - coumaric acid   ferulic acid

syringic acid   vanillic acid   p - hydroxybenzoic acid

Fig. 1   Phenolic allelochemicals (derivatives of cinnamic and benzoic acid)

Content of phenolic acids was expressed in µg/g dry weight (Table 1). Each sample was taken in 5 replicates.

Statistical analysis: The data were analyzed using t-test for dependent variables or when have large sample or more than two combinations one-way ANOVA.

3. EXTRACTION AND DETECTION OF TOTAL PHENOLICS AND PHENOLIC ACIDS IN PLANTS

The allelopathic effects of dominant plants on other plants in phytocoenosis are caused by phenolic phytotoxins present in all parts of plants, but the highest amount of these compounds is accumulated in the leaves. Leaves of dominant trees represent the main components of the litter in the forest, thus analysis of phenolic compounds and measurements of their content in leaves and leaf litter is considered as very important.

Experiment I. *Preparation of phenol-containing extracts*

Material required: Dry plant material, ethylacetate, methanol, anhydrous Na_2CO_3 and 2N HCl (HPLC grade), N_2, distilled water, graduated tube, glass flasks of 100 ml, funnels, separatory funnels, Soxlet apparatus, rotary evaporator, filter paper, shaker, mill, sieve of 0.5 mm in diameter, water bath, pipette of 1, 2 and 10 ml, test tubes of 20 ml.

Procedure: Both phenolic acids and total phenolic compounds were extracted from 5 x 2.0 g aliquots of dry plant material with 80% (v/v) boiling aqueous methanol solution (3x30 ml; 4 h) followed by ethylacetate (3x30 ml; 4 h) in glass flasks of 100 ml volume. The extraction was done in Soxlet apparatus under refluxing condition. After filtration pooled methanol and ethylacetate extracts were evaporated with a rotary evaporator under N_2, the residue was dissolved in 10 ml of distilled water adjusted to pH to 2.0 with 2N HCl and phenolics were transferred to ethylacetate (3x30 ml). The ethylacetate phase was dehydrated with anhydrous Na_2CO_3 and evaporated to dryness in a stream of nitrogen and the residue dissolved in 4 ml of 80% (v/v) MeOH. With this procedure, free phenolics (high soluble fraction) were prepared. Bound phenolics (fraction of phenolics either ester or ether linked to the cell walls polysaccharides, hemicelluloses or polymerized into lignin) were prepared by boiling the insoluble residue that remained after the first procedure in 10 ml of 2N HCl for 60 min (acid hydrolysis) and transferring to ethylacetate (3x30 ml). The ethylacetate phase was dehydrated with anhydrous Na_2CO_3 and evaporated to dryness in a stream of nitrogen using rotary evaporator and the residue dissolved in 4 ml of 80% (v/v) methanol solution. Both samples of free and bound phenolics were used for immediate HPLC analysis or stored at -20 °C until use.

Experiment II. *Determination of total phenolics spectrophotometrically*

Material required: Sample solution in methanol, distilled water, 20% Na_2CO_3, Folin-Ciocalteu's phenol reagent, water bath, graduated tube, cuvette, micro pipette (0.1 ml), pipette of 1, 2 and 10 ml, test tubes of 20 ml, pure ferulic acid (Serva, Germany), Shimadzu UV 160 spectrophotometer.

Procedure: A 0.02 ml aliquot of the sample solution for free phenolics and 0.005 ml for bound phenolics in methanol was taken and 7 ml distilled water plus 0.1 ml Folin-Ciocalteu's phenol reagent was added and after 3 min 0.2 ml of 20% Na_2CO_3 was added. After boiling at $90\,^0C$ (exactly 5 min) samples were cooled at room temperature and were diluted with H_2O to the volume of 10 ml. Only distilled water and reagents were used as a blank. The absorbance of total phenolics (free and bound) was measured at 660 nm spectrophotometrically (a Shimadzu UV 160 spectrophotometer), according to Feldman and Hanks (1968), with a sensitivity of 0.05 µg/g d.w. A standard curve was constructed with different concentrations of ferulic acid (Serva, Germany).

Statistical analysis: The data were analyzed using t-test for dependent variables or when have large sample or more than two combinations one-way ANOVA.

Precaution: When Shimadzu UV 160 spectrophotometer is used, the absorbance should not be higher than 0.430, because concentrations of ferulic acid for a standard curve is supposed to be between 0.33-80 µg/ml.

Experiment III. *Determination of phenolic acids by HPLC analysis*

Material required: Acetonitrile, o-phosphoric acid (HPLC grade), Nucleosil 100-5 C_{18} column (5 µm; 4.0x250 mm-Agilent Technologies, USA), vial, graduated tube, cuvette, distilled water, pure p-hydroxybenzoic and syringic acid (Acros Organics, U.S.A.), ferulic, vanillic and p-coumaric acid (Serva, Germany), HPLC (Hewlett Packard HP 1100).

Procedure: Phenolic acids were detected between 210 and 360 nm using a Hewlett Packard diode array detector (HP 1100 HPLC system). The separation was achieved with a Nucleosil 100-5 C_{18} column; 5 µm; 4.0x250 mm (Agilent Technologies, USA) at a flow rate of 1.0 ml/min and injection volume of 5 mL. For the elution, a discontinuous acetonitrile-water gradient was used: 15% acetonitrile (5 min), 30% acetonitrile (20 min), 40% acetonitrile (25 min), 60% acetonitrile (30 min), 60% acetonitrile (35 min) and

100% acetonitrile (45 min, isocratic). To avoid tailing of the phenolic acids, 0.05% *o*-phosphoric acid was added to the solvents. Phenolic acids were identified on the basis of retention time values and absorption peaks of pure *p*-hydroxybenzoic and syringic acid (Acros Organics, U.S.A.), ferulic, vanillic and *p*-coumaric acid (Serva, Germany) that served as references. Content of phenolic acids was expressed in μg/g dry weight.

Observations and data analysis: Each sample was taken in 5 replicates. The data are analyzed using t-test for dependent variables or when have large sample or more than two combinations one-way ANOVA.

4. DETERMINATION OF TOTAL PHENOLICS AND PHENOLIC ACIDS IN SOILS

During decomposition of plant remains, many phenolic compounds are released by leaching, microbial degradation or are synthesized by microbial activity. In forestry, problems of natural regeneration and reforestation are connected to the presence of phenolic substances deposited in the soil. Methods for extraction and identification of toxic substances from different soil types (mineral or organic) are described. The method for extracting of soil phytotoxins is based on the use of ethylacetate and methanol (free phenolics) and alkaline hydrolysis (bound phenolics).

Experiment I. *Extraction of phenolics from soil samples*

Materials required: Organic or mineral soil, ethylacetate, methanol, anhydrous Na_2CO_3, conc. HCl, 2N HCl, and 2N NaOH (HPLC grade), distilled water, N_2, graduated tube, flasks of 100 ml, filter paper, funnels, separatory funnels, Soxhlet apparatus, rotary evaporator, shaker, mill, sieve of 0.5 mm in diameter, water bath, pipette of 1, 2 and 10 ml, test tube of 20 ml, distilled water.

Procedure: After removal of visible plant remains, the soil was dried, milled and sifted through the sieve with 0.5 mm diameter holes. Free forms of phenolics were extracted from 2 g of dried organic soil (peat soil) or from 30 g of dried mineral soil with boiling ethylacetate (3x50ml, 12 h) and methanol (3x50ml, 12 h). The extraction was done in Soxet apparatus under refluxing condition. The extracts were evaporated in a stream of nitrogen, dissolved in water adjusting pH to 2.0 with 2N HCl and transferred to ethylacetate. The ethylacetate phase was dehydrated with anhydrous Na_2CO_3 and evaporated to dryness in a stream of nitrogen and the residue dissolved in 4 ml of 80% (v/v) MeOH (free phenolics). Residual soil was treated with 15 ml of 2N NaOH and after boiling for 24 h the mixture was acidified with concentrated HCl to pH 2.0, and filtered. This precipitate (humic acid fraction) was washed three times with 10 ml of water. Obtained supernatant was mixed with previous obtained supernatant (fulvo acid fraction). The combined supernatant was filtered through filter paper; the filtrate was extracted three times with ethyl acetate (3x50ml). The ethylacetate phase was dehydrated with anhydrous Na_2CO_3 and evaporated to dryness in a stream of nitrogen and the residue dissolved in 4 ml of 80% (v/v) methanol HPLC grade (bound phenolics). Both samples of free and bound phenolics were used for immediate HPLC analysis or stored at -20 °C until use.

Experiment II. *Determination of total phenolics spectrophotometrically*
Material required, Procedure, Statistical analysis and Precaution are the same as described in Section 3.2.

Experiment III. *Determination of phenolic acids by HPLC analysis*
Material required, Procedure, Statistical analysis and Precaution are the same as described in Section 3.3.

5. METHODS TO STUDY PHOTOSYNTHETIC EFFICIENCY, CHLOROPHYLL A AND B IN PLANTS

Phenolic compounds of dominant plants (donor plants) in plant community have inhibitory effects on photosynthesis of target plants. Methods of the photosynthetic efficiency (Fv/Fm) measurement, the extraction, detection and measurement of chlorophyll (a, and b) are described.

Experiment I. *Measurement of photosynthetic efficiency*

Principle: Chlorophyll fluorescence is a sensitive and early indicator of damage to photosynthesis and to the physiology of the plant resulting from the effect of allelochemicals, which directly or indirectly affects the function of photosystem II (Bolhar-Nordenkemf et al., 1989, Krause and Weiss 1991). This approach is convenient for a photosynthesis analysis *in situ* and *in vivo* and quick detection of otherwise invisible leaf damage. The photosynthetic plant efficiency was measured using the method of induced chlorophyll fluorescence kinetics of photosystem II [Fo, non-variable fluorescence; Fm, maximum fluorescence; Fv=Fm-Fo, variable fluorescence; $t_{1/2}$, half the time required to reach maximum fluorescence from Fo to Fm; and photosynthetic efficiency Fv/Fm].

Material required: Intact leaves and Portable Plant Stress Meter (BioMonitor S.C.I.AB, Sweden).

Procedure: Chlorophyll was excited for 2 sec by actinic light with a photon flux density of 200 mmol m^{-2} s^{-1}. Prior to measuring the chlorophyll,

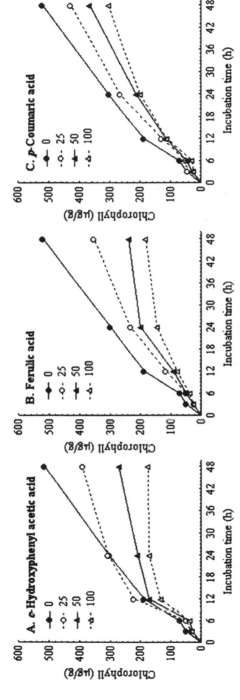

Fig. 2 The effect of three phenolic acids on the Chl content of 2-week-old rice seedling. A. o-Hydroxyphenyl acetic acid; B. ferulic acid; and C. p-Coumaric acid. Kimura's culture solution contains 0, 25, 50, or 100 ppm of either o-hydroxyphenyl acetic, ferulic or p-coumaric acids. Extraction was done on liquid-nitrogen frozen leaves with 80% acetone. The data is mean of three determinations and the bars indicate the standard deviation (Yang et al., 2002, 2004).

fluorescence samples were adapted to the dark for approximately 30 min to maximize the oxidation of the primary quinone electron acceptor pool of PS II and to enable the full relaxation of any rapidly recovering fluorescence quenching.

Observations: The apparatus allows the adjustment of excitation time to 4 sec or longer. Also, photon flux density of 200 μmol $m^{-2}s^{-1}$ can be elevated to 400 or 600 μmol $m^{-2} s^{-1}$, which depends on plant species or measuring conditions.

Statistical analysis: Each sample was taken in at least 5 replicates. The statistical analysis of the photosynthetic efficiency (Fv/Fm) of the species that were examined was performed using a one way-ANOVA analysis.

Experiment II. *Extraction and determination of chlorophyll a and b*

Material required: Leaf tissue, dimethyl sulphoxide, vial, graduated tube, cuvette, thermostat, pipette of 1, 2 and 10 ml, Shimadzu UV 160 spectrophotometer.

Procedure: The leaf disc (φ 1cm) was placed in a vial containing 1 ml dimethyl sulfoxide and chlorophyll was extracted at 60 ^{0}C by incubating for 30 min. After cooling the OD values at 645 and 663 nm were read in a Shimadzu UV 160 spectrophotometer. Chlorophyll a and b content was calculated following the equation used by Arnon (1949). The content of chlorophyll a and b was expressed in mg/g dry weight.

Observations and data analysis: If OD values are > 2.0, the extract must be diluted with dimethyl sulphoxide. Each sample was taken in 5 replicates. The data are analyzed using t-test for dependent variables or when have large sample or more than two combinations one-way ANOVA.

Precautions: If leaf samples are over mature (more cutinous) and thick, time of incubation may be prolonged.

6. BIOTESTS

Allelochemicals must be transmitted from the donor plant to the acceptor plant through either their rhizosphere soil or air (volatile components). These allelochemicals affect some plants and metabolic processes (seed germination, seedling growth, photosynthesis, respiration, ion uptake etc.). By using biotests, it is possible to prove that phenolics have no effect, or stimulate or inhibit a metabolic process. The biotests of seed germination and seedling growth of test plants in soil, the influence of aqueous extracts of leaves, bulbs, phenolic-containing extracts of surface layer (0-10 cm) of soil and volatile components of leaves and bulbs are described.

Experiment I. *Plant growth tests with soil*

Material required: Surface soil samples (0-10 cm), plastic dishes (500 ml), etiolated seedlings of test plants, pots and white fluorescent lamp.

Procedure: Test plants were grown in 500 ml plastic dishes. Surface samples (0-10 cm) of soil under *A. ursinum* (Wild garlic) as a dominant plant were collected in the middle of July after the withering of aboveground plant parts. Two-day-old etiolated seedling of lettuce (*Lactuca sativa*), amaranth (*Amaranthus caudatus*) and wheat (*Triticum aestivum*) were planted in experimental soil-containing pots and grown in a green-house for 15 d (14 h light, white fluorescent lamp 3.6 x10^3 erg/cm^2/sec, 25 °C and 10 h dark, 20 °C). The controls were grown under the same conditions in the pots containing forest soil taken from the spots without wild garlic. Measurements were performed using 50 individuals of each test plant species.

Observations: All experiments were repeated three times.

Statistical analysis: The data are analyzed using t-test for dependent variables or when have large sample or more than two combinations one-way ANOVA.

Experiment II. *Volatile compounds from leaves and bulbs*

Volatiles (terpenoids, ethylene and other compounds) can be released from the leaves and other plant parts and consequently can affect the germination of seeds, growth and development of neighbouring plants in ecosystems.

Material required: Seedlings of each test plant species, plastic dishes (200 ml), glass tanks (35x35x25 cm), fresh *A. ursinum* leaves or *A. ursinum* bulbs (50 g).

Procedure: Ten seedlings of each test plant species planted in the sand in plastic dishes (200 ml) in ten replicates were transferred into glass tanks (35x35x25 cm) containing 50 g *A. ursinum* leaves or 50 g *A. ursinum* bulbs which were replaced with fresh ones every 48 h. The controls were kept in empty tanks. The tanks were tightly covered with glass covers, transferred in a green-house at 25 °C and the plants were measured 10 d later.

Statistical analysis: All experiments were repeated three times. The data are analyzed using t-test for dependent variables or when have large sample or more than two combinations one-way ANOVA.

Experiment III. *Seedling growth tests*

Materials required: Test plant seeds, 5% sodium hypochlorite, distilled water, Petri dishes (10 cm dia), aqueous extracts of: (i) leaves, (ii) bulbs, (iii) phenol-containing extracts of surface layer of soil (0-10 cm).

Procedure: Test plant seeds were sterilized with 5% sodium hypochlorite for 10 min and thoroughly rinsed with distilled water, germinated at 25 °C in the dark for 36 h (lettuce), 48 h (amaranth) and 72 h (wheat). Uniform seedlings with hypocotyls or coleoptiles (approximately 5 mm long), were used. They were grown in Petri dishes (10 cm) containing 5 ml aqueous extracts of: (i) leaves, (ii) bulbs, (iii) phenol-containing extracts of surface layer of soil (0-10 cm) under *A. ursinum* (0.25 g/ml) or 5 ml distilled water (controls). Phenolic fraction can be obtained through the following procedure: 1.25 g of soil was extracted with 2x10 ml of ethylacetate (8 h), ethylacetate was evaporated to dryness and residue was dissolved in 5 ml distilled water. After 48 h growth at 25 °C, the length of radicle, hypocotyls or coleoptiles was measured. In each experiment, 50 seedlings were used and the experiments were repeated three times.

Statistical analysis: The data are analyzed using t-test for dependent variables or when have large sample or more than two combinations one-way ANOVA.

Experiment IV. *Seed germination tests*

Material required: Test plant seeds, 5% sodium hypochlorite, distilled water, Petri dishes (10 cm diameter), aqueous leaf or bulb *A. ursinum* extract (0.2 g fresh weight/ml), thermostat.

Procedure: Test plant seeds were sterilized with 5% sodium hypochlorite for 10 min and thoroughly rinsed with distilled water. The seeds (100 seeds of each) germinated at 25 °C in Petri dishes (10 cm diameter) containing 5 ml of aqueous leaf or bulb *A. ursinum* extract (0.2 g fresh weight/ml) or the same volume of phenolic-containing extract of surface layer of soil (0-10 cm) under *A. ursinum* (1.0 g dry soil/ml). Control seeds germinated in distilled water. Germinated seeds were counted after 24, 48 and 72 h.

Statistical analysis: The data are analyzed using t-test for dependent variables or when have large sample or more than two combinations one-way ANOVA.

Precautions: The experiments were performed in 20 replicates (each Petri dish contains 100 seeds). All experiments were repeated three times.

Observations: Plant seeds from examined phytocoenosis should be used.

7. REFERENCES

Arnon, D.I. (1949). Copper enzymes in isolated chloroplasts. Polyphenoloxidases in *Betula vulgaris. Plant Physiology* **24**: 1-15.

Bolhar-Nordenkampf, H.R., Long, S.P., Baker, N.R., Oquist, G., Schreiber, U. and Lechner, E.G. (1989). Chlorophyll fluorescence as a probe of the photosynthetic competence of leaves in the field: a review of current instrumentation. *Functional Ecology* 3: 497-514.

Chou, C.H. and Muller, C.H. (1972). Allelopathic mechanisms of *Arctostaphylos glandulosa* var. *zacaensis*. *American Midland Naturalist* 88: 324-347.

Djurdjević, L., Dinić, A., Mitrović, M., Pavlović, P. and Tešević, V. (2003). Phenolic acids distribution in a peat of the relict community with Serbian spruce in the Tara Mt. forest reserve (Serbia). *European Journal of Soil Biology* 39: 97-103.

Djurdjević, L., Dinić, A., Pavlović, P., Mitrović, M., Karadžić, B. and Tešević, V. (2004). Allelopathic potential of *Allium ursinum* L. *Biochemical Systematics and Ecology* 32: 533-544.

Einhellig, F.A., Schon, M.K. and Rasmussen, J.A. (1982). Synergistic effects of four cinnamic acids compounds on grain sorghum. *Journal of Plant Growth Regulation* 1: 251-258.

Feldman, A.W. and Hanks, R.W. (1968). Phenolic content in the roots and leaves of tolerant and susceptible citrus cultivars attacked by *Rodopholus similis*. *Phytochemistry* 7: 5-12.

Gallet, C., Nilsson, M.C. and Zackrisson, O. (1999). Phenolic metabolites of ecological significance in *Empetrum hermaphroditum* leaves and associated humus. *Plant and Soil* 210: 1-9.

Hennequin, J. R. and Juste, C. (1967). Presence of free phenolic acids in soil: Study of their influence on germination and growth of plants. *Annual Agronomy* 18: 545-569.

Hiscox, J.D. and Israelstam, G.F. (1979). A method for the extraction of chlorophyll from tissue without maceration. *Canadian Journal of Botany* 57: 1332-1334.

Ishikura, N. (1976). Seasonal changes in contents of phenolic compounds and sugar in *Rhus, Euonymus* and *Acer* leaves with special reference to anthocyanin formation in autumn. *Botanical Magazine*, Tokyo 89: 251-257.

Katase, T. (1981). The different forms in which *p*-coumaric acid exists in a peat soil. *Soil Science* 131: 271-275.

Katase, T. (1981). The different forms in which *p*-hydroxybenzoic, vanillic, and ferulic acids exist in a peat soil. *Soil Science* 132: 436-443.

Katase, T. (1983). The significance of hummification to different forms of *p*-coumaric and ferulic acids in a pond sediment. *Soil Science* 135: 151-155.

Katase, T. and Kondo, R. (1984). Distribution of different forms of some phenolic acids in peat soils in Hokkaido, Japan: 1. Trans-4-hydroxycinnamic acid. *Soil Science* 138: 220-225.

Khan, M.A. and Ungar, I.A. (1986). Inhibition of germination in *Atriplex triangularis* seeds by application of phenols and reversal of inhibition by growth regulators. *Botanical Gazette* 142: 148-151.

Kögel, I. (1986). Estimation and decomposition pattern of the lignin component in forest humus layers. *Soil Biology and Biochemistry* 18: 589-594.

Kögel, I. and Bochter, R. (1985). Characterization of lignin in forest humus layers by high-performance liquid chromatography of cupric oxide oxidation products. *Soil Biology and Biochemistry* **17**: 637-640.

Kögel-Knabner, I. (2002). The macromolecular organic composition of plant and microbial residues as inputs to soil organic matter. *Soil Biology and Biochemistry* **34**: 139-162.

Krause, G.H. and Weiss, E. (1991). Chlorophyll fluorescence and photosynthesis: the basics. *Annual Review of Plant Physiology* **42**: 313-349.

Lodhi, M.A.K. and Killingbeck, K.T. (1980). Allelopathic inhibition of nitrification and nitrifying bacteria in a ponderosa pine (*Pinus ponderosa* Dougl.) community. *American Journal of Botany* **67**: 1423-1429.

Lodhi, M.A.K. and Rice, E.L. (1971). Allelopathic effects of *Celtis laevigata*. *Bulletin of Torrey Botanical Club* **98**: 83-89.

Nagels, L. and Parmentier, F. (1976). Kinetic study of possible intermediates in the biosynthesis of chlorogenic acid in *Cestrum poeppigii*. *Phytochemistry* **15**: 703-706.

Oquist, G. and Wass, R. (1988). A portable, microprocessor operated instrument for measuring chlorophyll fluorescence kinetics in stress physiology. *Physiologia Plantarum* **73**: 211-217.

Rice, E.L. (1974). *Allelopathy*. Academic Press, New York, USA

Rice, E.L. (1979). Allelopathy-an update. *Botanical Review* **45**: 15-109.

Swain, T. (1976). Flavonoids. In: *Chemistry and Biochemistry of Plant Pigments* T. Goodwin (ed), pp. 166-206. Academic Press, New York, USA

Tsutsuki, K. and Kondo, R. (1995). Lignin-derived phenolic compounds in different types of peat profiles in Hokkaido, Japan. *Soil Science and Plant Nutrition* **41**: 515-527.

Tsutsuki, K., Kondo, R. and Shiraishi, H. (1993). Composition of lignin-degradation products, lipids and opal phytoliths in a peat profile accumulated since 32,000 years B.P. in Central Japan. *Soil Science and Plant Nutrition* **39**: 463-474.

Tsutsuki, K., Esaki, I. and Kuwatsuka, S. (1994). CuO-oxidation products of peat as a key to the analysis of the paleo-environmental changes in a wetland. *Soil Science and Plant Nutrition* **40**:107-116.

Whitmore, F.W. (1976). Binding of ferulic acid to cell walls by peroxidases of *Pinus elliottii*. *Phytochemistry* **15**: 375-378.

Yang, C.M., Lee, C. N. and Chou, C.H. (2002). Effects of three allelopathic phenolics on chlorophyll accumulation of rice (*Oryza sativa*) seedlings: I. Inhibition of supply orientation. *Botanical Bulletin of Academia Sinica* **43**: 299-304.

Yang, C.M., Chang, I.F., Lin, S.J. and Chou, C.H. (2004). Effects of three allelopathic phenolics on chlorophyll accumulation of rice (*Oryza sativa*) seedlings: II. Stimulation of consumption-orientation. *Botanical Bulletin of Academia Sinica* **45**: 119-125.

Chapter

Computational Methods to Study Properties of Allelochemicals and Modelling of Molecular Interactions in Allelopathy

E. Lo Piparo, P. Mazzatorta and E. Benfenati[†]

1. INTRODUCTION

In recent years, interest is increasing in sustainable agriculture, aimed at detailed and far-sighted evaluation of the effects of allelochemicals, that are widely used often in large amounts in the environment. Allelopathy has been evaluated as an alternative strategy for controlling weeds in particular but also insects and diseases (FATEALLCHEM EC project[1]). To know this large family of compounds using conventional experimental toxicological methods, would require huge investments in terms of both cost and time. In addition EU member states are determined to move away from animal studies in assessing the environmental safety of chemicals. Computational methods offer an attractive alternative to laborious and expensive experimental studies, with the potential for several studies on many compounds. Computational tools could evaluate numerous compounds, for a range of toxicological end-points. This would act as a 'screen', highlighting compounds for which experimental evaluation is necessary or recommendable and producing a system for predicting the toxicity of allelochemicals. This system can also give additional information, since it

Institute for Pharmacological Research "Mario Negri", via Eritrea 62, 20117, Milan, Italy.
E-mail: lopiparo@marionegri.it
[†]Corresponding author: E-mail: benfenati@marionegri.it, [1]http://www.fateallchem.dk/

can identify ecotoxic properties in a fast, reproducible and convenient way on the basis of the simple chemical structure.

Models to predict the activity and properties of chemicals based on the chemical structure had been studied since long and usually using the Quantitative Structure-Activity Relationship (QSAR), while the simpler expression Structure-Activity Relationship (SAR) refers to a qualitative approach. These models integrate knowledge from different fields (chemistry, toxicology and computer science) that together can show a possible relationship to predict the property of a chemical from its structure, without any direct experimental measurement. In the past decade, a lot of effort has been made to develop Quantitative Structure-Activity Relationship models and methods because they can reduce the amount of resources and time needed to evaluate the chemicals to be processed. They are now attracting the attention of regulators as support tools in the assessment of several properties of chemicals. In this case, QSAR may offer an alternative for potentially more hazardous compounds to be studied.

Progress in biomolecular X-ray crystallography continues to indicate a number of important proteins as targets for bioactive agents in the control of animal and plant diseases, or simply as a key to understand fundamental aspects of their action. The docking method was developed as an automated procedure for studying and predicting the interaction of ligands with biomacromolecular targets. Using protein modelling together with molecular biology, the identification of biological processes can be accelerated, to predict interactions between molecular complexes as well as the strength of binding and to better understand the mechanism of action.

2. QSAR APPROACH TO PREDICT THE TOXICITY OF ALLELOCHEMICALS

Quantitative Structure-Activity Relationship (QSAR) approach was first developed by Cros (1863) and Brown and Fraser (1868). In the 1960s, C. Hansch, T. Fujita, S. M. Free Jr. and J. W. Wilson started what is now considered to be classical QSAR. A series of powerful advanced computer tools have now been introduced, increasing the capacity of QSAR.

Quantitative Structure-Activity Relationship studies search for a relationship between the activity/toxicity of chemicals and the numerical representation of their structure and/or features. The overall task is not easy. For instance, several environmental properties are relatively easy to model, but some toxicity endpoints are quite difficult, because the toxicity is the result of many processes, involving different mechanisms. Toxicity data are also affected by experimental errors and their availability is limited because experiments are expensive. A 3D-QSAR model reflects the characteristics of chemicals based on their three-dimensional organization in space, so the structure is a key factor. This kind of approach is made up of several steps:

(i) Build-up of a data set containing chemicals within a specified, well-defined end-point,
(ii) Evaluation of the minimum energy conformation for each compound (optimization of geometry),
(iii) Calculation of descriptors for each chemical,
(iv) Find the relationship between activity and descriptors by some statistical technique that gives a variable selection to ascertain the best predictive model.

The overall strategy of the molecular modelling followed is shown in Figure 1.

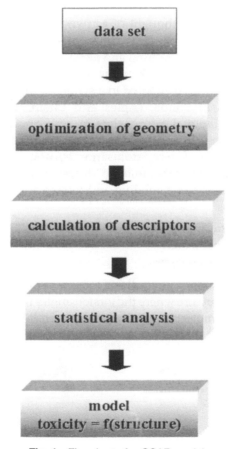

Fig. 1 Flowchart of a QSAR model.

2.1 Toxicity Data

To produce QSAR models, a data set containing chemicals within a specified well-defined end-point is necessary. Since our knowledge about the properties of the natural compounds that surround us is very poor, especially for allelochemicals and toxicological evaluation of synthetic pesticides is well documented (regulators oblige the chemical industry to produce experimental data for synthetic chemicals, before they can be marketed), when allelochemicals' toxicity values are not available, pesticides with similar structure can be used in the analysis. Therefore suitable data sets can be defined with pesticides and their activities, to predict the toxicity (activity) of the allelochemicals.

2.2 Chemical Structures and Optimization of Geometry

Chemicals are commonly thought of as two-dimensional structures, but their toxic effects are an expression of their three-dimensional structure. To correctly describe the 3D structural and electronic properties of a molecule under study in Quantitative Structure-Activity Relationship (QSAR), typically the structure is considered in a stable (optimized) state, in which the internal strain of bonds, angles and torsion angles is minimal. Theoretical-computational approaches in conformational searches are based on the degrees of freedom in the molecule. Even a small molecule composed of a few tens of heavy atoms (non-hydrogen atoms) containing a few rotatable bonds has a high number of degrees of freedom, which include bond stretching, bond angle bending and torsion angle rotations. For this there are three steps:

  (i) Building from 2D sketches
 (ii) Conformational search
(iii) Optimization of geometry

In Figure 2 an example is given: the 2D structure of DIMBOA is converted in a optimized 3D structure where the torsion angles have the biggest impact on the energy of the molecule and dictate the overall 3D molecular shape.

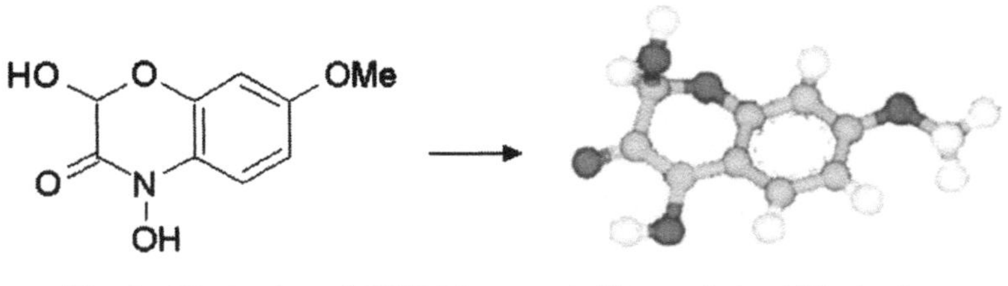

Fig. 2 2D structure of DIMBOA converted in a optimized 3D structure.

Moreover Figure 3 shows the graph of the potential energy of the molecule where E2 and E4 representing structure optimization finished in a local minimum and E6 a global minimum found (internal strain of the molecule minimized).

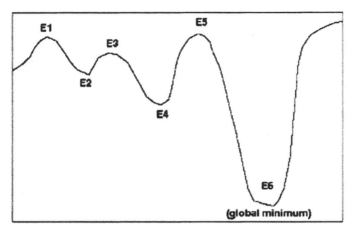

Fig. 3 Graphic of the potential energy of the molecule, where E2 and E4 representing structure optimization finished in a local minimum and E6 a global minimum found.

3. CHEMICAL DESCRIPTORS

When suitable data sets are defined, different approaches can be used to codify the chemical information within chemical descriptors. Nowadays we have powerful tools to describe them in different ways by their physicochemical properties, surface properties, or their 3D fields generated by interactions with different chemical probes. Typically many chemical descriptors are calculated (up to thousands) and then the important ones are selected. We briefly explain some of the most common approaches used, trying to classify them in 'families' to simplify the overview.

3.1 Molecular-based Descriptors

These only describe the magnitude of particular physical properties but no directional preferences in relation to predefined oriented axes: the number of a specified atom, molecular fragments or functional groups; molecular indices obtained by topological methods (e.g. molecular connectivity indices related to the degree of branching in the compounds), atomic properties (e.g. electrotopological indices, atomic polarizability), geometric properties (e.g. molecular surface area, volume, moment of inertia, shadow area, projections and gravitational indices), electrostatic properties (e.g. partial atomic charges), physicochemical properties (e.g. partition coefficient between octanol and water). In Table 1 the values of some molecular-based descriptors for BOA (benzoxazolin-2-one),

Table 1   Calculation of some molecular-based descriptors for BOA, DIMBOA and MBOA. Physicochemical descriptor like logP (partition coefficient between octanol and water); constitutional descriptors like the number of a specified atoms or bonds (number of carbons, hydrogens, oxygens, nitrogens, single and aromatic bonds, the total number of atoms and bonds) and molecular weight; quantum-mechanical descriptors like HOMO (Highest Occupied Molecular Orbital) and LUMO (Lowest Unoccupied Molecular Orbital).

Molecule	BOA	DIMBOA	MBOA
logP	1.56	0.28	1.35
Atoms	15	24	19
C	7	9	8
H	5	9	7
O	2	5	3
N	1	1	1
Bonds	16	25	20
Single bonds	9	18	13
Aromatic bonds	6	6	6
Rings	2	2	2
HOMO	-0.34229	-0.32604	-0.33076
LUMO	-0.00681	-0.00344	-0.0065

DIMBOA (2,4-dihydroxy-7-methoxy-1,4-benzoxazin-3-one) and MBOA (6-methoxy-benzoxazolin-2-one) are showed as an example: physicochemical descriptor like logP (partition coefficient between octanol and water); constitutional descriptors like the number of a specified atom or bond (number of carbons, hydrogens, oxygens, nitrogens, single and aromatic bonds, total number of atoms and bonds) and molecular weight; quantum-mechanical descriptors like HOMO (Highest Occupied Molecular Orbital) and LUMO (Lowest Unoccupied Molecular Orbital).

3.2   Comparative Molecular Field Analysis (COMFA)

It describes the properties by fields (steric and electrostatic) calculated in a regular grid (micro-environment surrounding the molecules) using small probes. Figure 4 schematises the Calculation of Comparative Molecular Field Analysis (CoMFA) descriptors: at each grid point, the steric and electrostatic interaction energies between the probe and each molecule are evaluated and recorded, producing a 3D box of interaction energies that become new steric and electrostatic descriptors in a QSAR analysis.

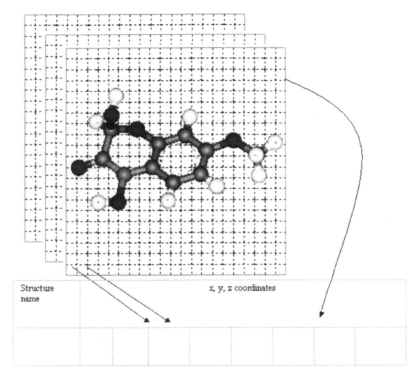

Fig. 4 Calculation of Comparative Molecular Field Analysis (CoMFA) descriptors. This technique measures the steric and electrostatic interaction energies between a small probe at a series of regular grid positions around the molecules. At each grid point, the steric and electrostatic interaction energies between the probe and each molecule are evaluated and recorded, producing a 3D box of interaction energies that become new steric and electrostatic descriptors in a QSAR analysis

As CoMFA is a 3D approach using grid point interaction energies as descriptors, before analysis the structures have to be aligned to share the 3D spatial arrangement of the major functional groups. Rational alignment is achieved by superimposing the molecules, according to orientation rules which follow from the common skeletons. Only then the spatial regions that are strongly related to the activity can be discovered. The advantages of CoMFA methods are that describing properties in terms of 3D fields, means the results can be visualized by a 3D space contour map of favourable and unfavourable regions of the different fields and one can locate points within the spatial distribution of properties that are strongly related to the activity. This visual inspection makes it easier to decide in which parts of the molecules, certain structural changes are sensitive to the biological activity. In Figure 5 an example of steric CoMFA contour map is given: this CoMFA steric contour map indicates that there is a small steric favourable area

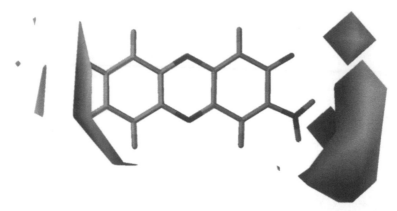

Fig. 5 Example of steric CoMFA contour map results using APO (2-amino-3H-phenoxazin-3-one) for visualization: near green area greater toxicity is correlated with more bulky groups and near yellow area with the less bulky group.

(green) in which the presence of bulky functional groups increases the toxicity and two steric unfavourable (yellow) areas correlated with the less bulky group. The major problem and most difficult steps of CoMFA arise from the alignment of compounds, which makes it difficult to study very heterogeneous data sets.

3.3 Field-fit Alignment (Seal Program)

Today it is an alternative method for alignment of molecular structures that maximizes the steric and electrostatic overlap using randomly generated starting configurations and keeping only the best results based on the value of the alignment function.

3.4 GRIND (GRid INdependent Descriptors)

3D-molecular descriptors, alignment-independent and based on molecular interaction, called GRIND have been developed. These are autocorrelation transforms that are independent of the orientation of the molecules in 3D space. The original descriptors can be extracted from the autocorrelation transform with the ALMOND program. The basic idea is to compress the information present in 3D maps into a few 2D numerical descriptors which are very simple to understand and interpret.

4. STATISTICAL ANALYSIS

The target is to apply a statistical method to correlate the descriptors with the biological activities. Partly due to the ease with which a variety of theoretical descriptors can be generated, QSAR researchers are often confronted with high-dimensional data sets; the task in such a situation is to solve an ill-posed problem in which there are more variables (descriptors) than objects (compounds). The inflation of parameters poses the problem of variable selection because the large number of variables in the model greatly increases the risk of chance correlation. The situation is further complicated because the underlying physicochemical attributes of the molecules that are correlated with their biological activities are often not known, so *a priori* feature selection is not feasible in most cases. There are two major advantages of feature selection. First, it can help define a model that can be interpreted. Second, the reduced model is often more predictive, partly because of the better signal-to-noise ratio, which is a consequence of pruning the non-informative inputs. In the past, variable or feature selection was done by a human expert who relied on experience and scientific intuition. However, when the dimensions of the data are huge, and the relations between variables are convoluted, human judgment may be inadequate. Recent developments in computer science have let to the creation of intelligent algorithms capable of finding optimal, or near-optimal, solutions for this problem.

Having selected the major features, the final stage of QSAR model building involves a feature mapping procedure.

Among the numerous methods used to formulate a mathematical relationship the following are prime examples.

4.1 Linear Methods

Multiple linear regression (MLR) analysis was the traditional approach for QSAR applications in the past. The major advantage of this method is its computational simplicity, so the resulting equation can be interpreted easily. However, this method becomes inapplicable as soon as the number of input variables equals or exceeds the number of observed objects. As a thumb rule, the ratio of objects to variables should be at least five for MLR analysis; otherwise there is a correspondingly large risk of chance correlation. A common way to reduce the number of inputs to MLR, without explicit feature selection, is through feature extraction by principal component analysis (PCA). The complete set of input descriptors is transformed to its orthogonal principal components, relatively few of which may suffice to capture the essential variance of the original data. The new principal components are then used as the input to a regression analysis. Another very powerful multivariate statistical method for an under-determined data set is partial least squares (PLS). Briefly, PLS attempts to identify a few latent structures, or linear combinations of descriptors, that best correlate with the observations. Unlike MLR, there is no restriction in PLS on the ratio of data objects to variables and PLS can deal with strongly co-linear input data and tolerates some missing values.

4.2 Non-Linear Methods

Traditionally, non-linear correlations in the data are explicitly dealt with by a predetermined functional transformation before entering a MLR. Unfortunately, the introduction of non-linear or cross-product terms in a regression equation often requires knowledge which is not available *a priori*. Moreover, it adds to the complexity of the problem and often leads to insignificant improvement in the resulting QSAR. To overcome this deficiency of linear regression, there is increasing interest in techniques that are intrinsically non-linear. At present, artificial neural networks (ANN) are probably the most widely used non-linear methods in chemometric and QSAR applications. ANNs are computer-based simulations that mimic the biological nervous systems. As in nature, they are composed of simple processing elements (neurons) operating in parallel (layers). The network function is determined largely by the connections between elements (weights), which are responsible for the network's intelligence. The inter- and intra-layer connections define the architecture of the ANN. ANNs can be trained to fit a particular function by adjusting the values of the weights between neurons. Supervised learning is a recursive learning process where inputs are fed into the ANN and mapped in the output; the output is then compared with the target and network weights are adjusted accordingly until the network output matches the target. A multiple-layer neural network can approximate a continuous function to an arbitrary accuracy, given a large number of neurons. On the other hand, unsupervised learning does not need target values, but it learns to recognize similarity among inputs.

Hunger and Hansch (1973) stated "One must rely heavily on statistics in formulating a quantitative model but, at each critical step in constructing the model, one must set aside statistics and ask questions. [...] without a qualitative perspective one is apt to generate statistical unicorns, beasts that exist on paper but not in reality. [...] it has recently become all too clear that one can correlate a set of dependent variables using random numbers as dependent variables. Such correlations meet the usual criteria of high significance". As such, model validation is a critical, but often neglected, component of QSAR development. In a recent review, Kövesdi et al state that "[...] in many respects, a proper validation process is more important than a proper training. It is all too easy to get a very small error on the training set, due to the enormous fitting ability of the neural network and then one may erroneously conclude the network would perform excellently".

4.3 Validation

The first benchmark of a QSAR model is usually to determine the accuracy of the fit to the training data. However, because QSAR models are often used for predicting the activity of compounds that have not yet been synthesized, the most important statistical measures are those giving an indication of their prediction accuracy. Common methods to test QSAR predictivity are listed below.

(i) Cross valid action: The most popular procedure for the estimation of the prediction accuracy is cross-validation, which includes techniques such as jack-knife, leave-one-out (LOO), leave-group-out (LGO) and bootstrap analyses. This group of methods is based on data splitting, where the original data set is randomly divided into two subsets. The first is a set of training compounds used for exploration and model building and the second is called the validation set for prediction and model validation. The procedure can be repeated several times, so that a set of predicted values can be obtained.

(ii) Y-Scrambling Test: It is another popular means of statistical validation. In this procedure the output values, i.e. biological responses, of the compounds are shuffled randomly and the scrambled data set is correlated by the QSAR method with the original X variables block. The entire procedure is repeated several times on differently scrambled data sets. If a strong correlation remains between the descriptors selected and the randomized response variables, then the significance of the proposed QSAR model is regarded as suspect.

(iii) External Test: The real criterion for the validation of a QSAR model can only be good predictivity for an external test set model has never seen before. The compounds in the external test set must not be used in any manner during the model building process. Otherwise the introduction of bias from the test set compromises the validation. Of course, the chemical space of the training and test sets must not be too different.

A variety of statistical parameters have been reported in the QSAR literature to reflect the quality of the model. These measures give indications about how well the model fits existing data, i.e., they measure the explained variance of the target parameter y in the biological data. Some of the most common measures of regression are root mean squares error (rmse), standard error of estimates (s), and coefficient of determination (R^2).

Generally, when the cross-validation performance is significantly better (> 0.5) than that of y-scrambling tests, but not very different from the training set and external test predictions, it is regarded as a good trait of a robust, high-quality QSAR model.

Examples of statistical analysis for QSAR developing using different techniques is given in Table 2 that shows the prediction of the *Daphnia magna* toxicity for allelochemicals (BOA, DIMBOA and MBOA) using QSARs models obtained with different statistical techniques (PLS, MLR, and Neural Networks).

Table 2 Prediction of the Daphnia Magna toxicity for allelochemicals (BOA, DIMBOA and MBOA) using QSARs obtained with different statistical techniques (PLS, MLR, and Neural Networks).

Allelochemical	PLS	MLR	NN
BOA	1.17	0.3	0.84
DIMBOA	2.25	0.8	0.92
MBOA	1.8	1	0.80

5. BIOMOLECULAR INTERACTIONS IN ALLELOPATHY

Biomolecular interactions are the core of all regulatory and metabolic processes, that together constitute the process of life. Computer-aided analysis of these interactions is becoming increasingly important as the number of known biomolecular structures grows and increasing processing power makes the analysis and prediction of molecular interactions easier. Protein modelling can be very important in structure determination, providing a computational tool to assist researchers in the determination of biomolecular complexes.

To investigate the complex relationship between biological activity and molecular structure the concept of molecular similarity is very important, because compounds with similar physicochemical properties can be expected to have similar biological activity. Sometimes chemical similarity cannot be defined in terms of overall properties, size and shape or certain structural features, but molecules may be similar with respect to certain target proteins, but dissimilar with respect to others, depending on the problem under study. Till we do not know the structure of the binding site, it is difficult to establish an objective similarity measure to predict biological activity. For this reason protein modelling has been recently integrated with QSAR, obtaining very important and helpful additional information: binding affinity can be included like chemical descriptors, classification of the compounds (in terms of receptor affinity), or using knowledge about the interaction with the binding site to align the molecule when it is required (e.g. CoMFA analysis).

The cytochrome P_{450} (CYP) families comprise the main enzymes involved in Phase I metabolism of foreign compounds, and their many forms show specificity in the metabolism of different substrates and give metabolic products with different toxicity rates. CYP1A2 bioactivates several substrates such as lipophilic polyaromatic/heteroaromatic molecules, with a structure similar to or the same as the allelochemical. We can therefore expect essential metabolic changes to contribute to final toxicity. Cytochrome P_{450} has an important role in the metabolisation of pesticides, with a structure similar to allelochemicals. Therefore cytochrome P_{450} docking can be used to investigate the chemical processes of these compounds.

5.1 Docking (Modelling Receptor Ligand Interaction)

Crystallography contributes significantly to biostructural research and gives many details about the structure and function of macromolecules. But often data are available for the shape of a protein and a ligand separately and not for the two together. Docking is the process in which two molecules fit together in 3D space and predicts how small molecules (ligand) bind to a receptor of known 3D structure in an energetically favourable way.

The techniques for automated docking fall into two categories:

I. Matching methods: They create a model of the active site and then attempt to dock a given molecule structure by matching its geometry to that of the active site.

II. Docking simulation methods: These are slower than the matching one, study the ligand outside the protein and randomly explore translations, orientations and conformations until an ideal site is found.

When the ligand is placed or found inside the receptor pocket, then the free energy of binding of the molecular complex is estimated computationally. Therefore the 3D-coordinates of the atoms in the protein receptor, a structural formula of the ligand, with bond lengths and angles and in addition knowledge of the position of the active site are required.

The generally accepted method when the crystallographic structure of the target protein is not available, involves sequence homology alignment with those isozymes for which crystallographic data are known (an appropriate template is a protein of known structure that shares a minimum of 20-25% homology in the amino-acid sequence), followed by amino acid residue replacement, insertion and deletion, as required by the alignment and finally, energy minimization of the raw structure.

As shown in Figure 6, the solvent molecules tend to be ordered around the molecules and when the protein and the ligand bind, several of these molecules are liberated and become disordered (entropic effect). Therefore, upon complex formation water molecules are released, receptor and ligand lose degrees of freedom and the interaction between the ligand and the receptor is calculated.

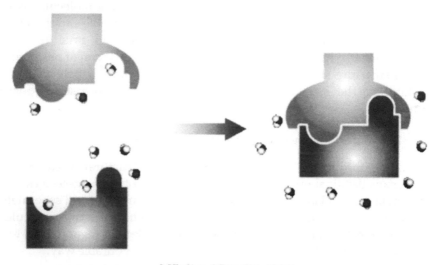

Affinity: $\Delta G = \Delta H - T\Delta S$

Fig. 6   Scheme for the binding of a receptor and a ligand in the solvated phase. Ligand bind, several of these molecules are liberated and become disordered (entropic effect). Therefore, upon complex formation water molecules are released, receptor and ligand lose degrees of freedom and the interaction between the ligand and the receptor is calculated.

The primary question in all docking programmes trying to address is what combination of orientation and conformation (pose) is the most favourable, so sampling is done across the entire range of positional, orientational and conformational possibilities. The task is to generate a set of energetically favourable configurations and stabilizing molecular interactions gives an estimate of the free energy of binding of this complex.

In any docking scheme, two conflicting requirements must be balanced: (i) the desire for a robust and accurate procedure and (ii) reasonable run time of the laboratory workstation for the computational method chosen.

The ideal procedure would find the global minimum in the interaction energy between the substrate and the target protein, exploring all variable degrees of freedom for the system. To meet these demands a number of docking techniques simplify the procedure and the user will choose the ones closest to the topic being studied.

6. SUGGESTED READINGS

Cramer, R.D., Patterson, D.E. and Bunce, J.D. (1988). Comparative molecular field analysis (CoMFA). Effect of shape on binding of steroids to carrier proteins *Journal of American Chemical Society* **110**: 5959-5967.

Cronin, M.T.D., Jaworska, J.S., Walker, J.D., Comber, M.H.I., Watts, C.D. and Worth, A.P. (2003). Use of QSARs in international decision-making frameworks to predict health of chemical substances. *Environmental Health Perspective* **111**: 1376-1401.

Eriksson, L., Jaworska, J., Worth, A.P., Cronin, M.T.D. and McDowell, R.M. (2003). Methods for reliability and uncertainty assessment and applicability evaluations of classification- and regression-based QSARs. *Environmental Health Perspective* **111** : 1361-1375.

Jaworska, J.S., Bomber, M., Auer, C. and Van Leeuwen, C.J. (2003). Summary of a workshop on regulatory acceptance of (Q)SARs for human health and environmental endpoints. *Environmental Health Perspective* **111** : 1358-1360.

Karelson, M. and Lobanov, V.S. (1996). Quantum-chemical descriptors in QSAR/ QSPR studies. *Chemical Reviews* **96**: 1027-1043.

Kearsley, S.K. and Smith, G.M. (1990). An alternative method for the alignment of molecular structures: Maximizing electrostatic and steric overlap. *Tetrahedron Computer Methodology* **3**: 615-633.

Klebe, G. (1998). *Comparative Molecular Similarity Indices: CoMSIA* In 3D QSAR in Drug Design. ., H. Kubinyi, G. Folkers, and Y.C. Martin (eds). Kluwer Academic Publishers, London, UK. **3**: 87.

Lo Piparo E., Fratev, F., Mazzatorta, P., Smiesko, M., Fritz, J.I., Benfenati, E. (2006). QSAR Models for *Daphnia magna* toxicity prediction of Benzoxazinone allelochemicals and their transformation products. *Journal of Agricultural and Food Chemistry* **54**: 1111-1115.

Lo Piparo, E., Smiesko, M., Mazzatorta, P., Jacqueline Indinger, Sylvia Bluemel, Benfenati, E. (2006). Preliminary analysis of toxicity of Benzoxazinones and their metabolites for *Folsomia candida*. **54**: 1090-1104.

Morris, G.M., Goodsell, D.S., Halliday, R.S., Huey, R., Hart, W.E., Belew, R.K. and Olson, A.J. (1998). Automated docking using a lamarckian genetic algorithm and empirical binding free energy function. *Journal of Computational Chemistry* **19**: 1639-1662.

Pastor, M., Cruciani, G., McLay, I., Pickett, S. and Clementi, S. (2000). GRid-INdependent descriptors (GRIND): a novel class of alignment-independent three-dimensional molecular descriptors *Journal of Medicnal Chemistry* **43**: 3233-3243.

Todeschini, R. and Consonni, V. (2000). *Handbook of Molecular Descriptors*, R. Mannhold, H. Kubinyi, H. Timmerman (eds) Vol. **11**. *Methods and Principles in Medicinal Chemistry*. Wiley-VCH, Weinheim, Germany.

Allelopathic Pollen: Isolating the Allelopathic Effects

S.D. Murphy

1. INTRODUCTION

Pollen allelopathy is one consequence of heterospecific pollen transfer and occurs only if sufficient numbers of grains containing exuded or surfacial allelochemicals alight on the stigma or other tissues of recipient individuals. Pollen allelopathy is not particularly cryptic but it is uncommon enough that it has not been as extensively studied as other forms of allelopathic interactions, with the most interesting recent findings coming from Victoria Roshchina's laboratory. At this juncture, some progress has been made in the crude and refined extraction of pollen allelochemicals and the mode of action of pollen allelopathy.

2. MEASURING POLLEN TRANSFER

While more commonly used to count or otherwise characterize cells for medical applications, Coulter Counters and flow cytometry technique can also be applied to the analysis of pollen grains in allelopathic studies. They are quite useful in determining the size and number of pollen grains. The technique is often used for assessing the production and size of pollen from the originating individual rather than how much was transferred to heterospecific stigma, as would be needed in a basic assessment of potential allelopathic interactions.

Department of Environment and Resource Studies University of Waterloo, Waterloo, Ontario, N2L 3G1, Canada   Phone: 519-888-4567 x5616, Fax: 519-746-0292. E-mail: sd2murph@fes.uwaterloo.ca

2.1 Coulter Counters

Principle: A Coulter Counter helps to determine the size and number of pollen grains, though the only major published work on using Coulters for intact pollen appears to be of Phamdelegue et al. (1994). Most of the principles of the Coulter Counter are available via a simple search of the World Wide Web; A summary is provided based on the author's interpretation after using this type of device for over a decade . A finger tube containing pollen is immersed (but only partially) in a reservoir. A hole in the tube allows a full tube and its contents to flow through the hole to equalize the volumes; this allows pollen to enter the Counter. Platinum electrodes (that will not electrolyze in solutions) in the finger and reservoir creates an electrical current that passes through the tube hole; two electrodes in the finger measure a known volume of liquid before the liquid and its particle move (this is calibrated, based on the spacing between electrodes). When more fluid with pollen is added above the upper sensor, this causes the fluid to drain through the tube hole. Once the meniscus is no longer in contact with this sensor, the Counter starts. When the meniscus is no longer in contact with the lower tube sensor, the Counter stops as all detectable fluid has drained through the tube's hole (some is still in the tube but the sensor is physically above this volume). The automatic detection means one has a record of the number and size of pollen grains (in this case) that were 'injected' in the sample volume Basically, the Counter is measuring the sudden increase in resistance (disruption of the current between the finger and reservoir) as a pollen grain passes through the tube hole.

Materials required: Acetone solution or other polar solutions, standard saline solution of Isoton II, commercial products Multisizer II or 3 (Beckman Coulter Ltd, United Kingdom), Beckman Coulter Ltd or others similar technique.

Procedure of pollen preparation: Pollen can be washed off stigmas with an acetone solution as water or other polar solutions often fail to sufficiently break electrostatic bonds holding heterospecific pollen to stigma. However, this means the acetone must be evaporated in an air drying oven (48 h) because a Coulter Counter requires a saline solution of standard volume (usually 10-20 ml) be used to prepare pollen samples. If the solutions are mixed and the volumes are inconsistent, there is a risk that differences in conductivity will create errors.

It requires a lot of pollen but comparing samples of masses of pollen measured on a microbalance and the effect of using a standard saline solution of Isoton II after using acetone and (i). then drying it for 48 h, (ii). not drying it but waiting 48 h, or (iii). not drying it and processing immediately.

Using either of the commercial products Multisizer II or 3 (Beckman Coulter Ltd, United Kingdom), the results are similar. If the acetone is not dried, the results can vary by an order of magnitude in extreme cases and generally make the counts unreliable. 48 h may appear to be arbitrary but precisely how long is really required is yet to be tested.

Advantages and disadvantages: The most recent version of the Coulter Counter is able to differentiate between conspecific and heterospecific pollen grains that differ in diameter and volume at the μm scale. This, however, depends on the putatively allelopathic heterospecific pollen having a significantly different and detectable diameter and volume versus conspecific pollen and non-allelopathic heterospecific pollen. This is not always the case, hence the researcher must first a measure or otherwise know the diameter and volume differences between pollen of different species that are likely or could alight on stigmas of species vulnerable to heterospecific pollen transfer. Coulter Counters themselves are quite useful for determining volumes of pollen as a researcher could first determine this by measuring the cardinal dimensions of pollen grains under a compound microscope with a vernier stage and using MathCAD or similar programs to fit polynomials to determine the volume of the typically irregular (though often close to ovoid or ellipsoid) grains.

2.2 Flow Cytometry

Flow cytometry is an alternative for the studies of pollen allelopathy that offers a more accurate census and a wider range of sizes of pollen that can be analyzed, though it is much more expensive, in terms of the cost and operation of the equipment. It is in fact expensive enough (typically several hundred thousand dollars for the equipment), that the idea has been vetted but few were willing to risk the expense for simply counting as opposed to more sophisticated uses to investigate cellular physiology.

Principle: In its most basic description (Ormerod 2000, Shapiro 2003), flow cytometry uses a laser that is projected through a liquid injection of cells. When the wavelengths of light in the laser collide with the cells, each component of the cell or type of cell will give off unique signals that are (typically) converted into numerical values so data may be analyzed. Generally, cells are stained using fluorescent dyes bound to specific cytosolic structures or chemicals. The laser will 'excite' the dyes and cause them to emit longer-wavelengths of light measured by the flow cytometer. Many cytometers can sort cells by using the emission signals as comparative criteria to induce an electrical charge that diverts each cell of interest – each cell is contained within a droplet within the liquid injection stream and the charge can divert this droplet, drop it through a duo of metal plates that attract the droplets with opposite polarity and into a collection vessel or even a slide.

Materials required: As described in the principles, a researcher needs access to a flow cytometer plus fluorescent dyes; generally, these are sold as intact units, though the fluids used in the injection wells can differ - but they should not unduly alter the current between the electrodes lest confounding effects occur.

Procedure: Though expensive (because of laser and other precision instruments run by high powered computers), some species of fungi have already been examined with flow cytometry. Unlike pollen, it is more difficult to use flow cytometry because fungi are generally not stainable and this is needed for cytometry to work successfully. Nonetheless, both alighted and airborne conidia have been examined. In terms of pollen allelopathy, the alighted propagules are more important but it would be useful to determine the actual airborne concentrations of heterospecific pollen on a diurnal basis and this method would be a breakthrough. One real problem for fungi and, by extension, for pollen is to separate the structure of interest from assorted debris, e.g. in analyzing for heterospecific pollen transfer this would include parts of the stigma that can wash off. However, if the samples to be run through the cytometer are 'calibrated' by multiparametric comparisons of size, granularity and fluorescence, then the cytometer can distinguish fungi or pollen from debris.

What remains to be determined in the application for pollen, is whether the calibration is sensitive enough to distinguish between pollen of different species. To date, no one has explicitly tested mixed samples of more than two species and there were still difficulties isolating conidia from abiotic debris. Much like the issue for Coulter Counters, discriminating pollen will depend on which species are phenologically concurrent; in the case of fungi, this might be in thousands but for higher plants it generally will be in tens and many of these will be species that either flower at different times of the day or night and/or ones that have different pollination systems that preclude transfer (e.g. animal vs. wind pollination).

3. MEDIA FOR POLLEN GERMINATION

Just as a Coulter Counter or a flow cytometer can 'automate' or at least increase efficiency in counting heterospecific pollen transfer, use of artificial germination media still is an effective method of screening for pollen allelopathy. The fundamental approach recommended has not changed much in the last decade, i.e. sequentially dilute extracts from non-macerated pollen of putative allelopathic species and add these extracts to artificial media. For most species, especially trinucleate pollen as in Poaceae, it is important that the media be spread thinly to avoid anoxic wells from forming and that the media be contained in a Petri dish or another device that allows for high relative humidites for rehydration of pollen on the media. This is normally accomplished simply by paraffin-sealing a dish with media containing nutrients and any allelopathic extracts and pollen from species likely to be vulnerable to any allelopathic interactions.

Principle: The challenge has always been to provide a media suitable for pollen of most species, as it reduces preparation time if it can be mass produced. The additional issue is that the media must not create undue risks of contamination by fungi, bacteria, or other biota. Even though the dishes and liquid phase media may be autoclaved (before any extracts or pollen are introduced, of course) and sterile techniques used, contamination is always a risk. To reduce this contamination risk and provide a generic medium for pollen germination, the best approach is to reduce the use of sucrose by substituting other carbohydrates like maltose, saccharose, lactose, or raffinose as these are not easily metabolized by many microorganisms found as contaminants in growth media. For most species, with an emphasis on the more difficult to germinate trinucleate pollen, raffinose appears to be metabolized most effectively and promotes pollen germination, e.g. 81.7% germination in pollen from *Triticum aestivum*.

Recommended media: After several years of trying different concentrations of chemicals that have a large scientific pedigree at least to the innovations of Brubacker and Kwak, the author's results indicate the following media will germinate pollen at success rates of 75-90%, a figure that varies with individuals and species as varying proportions of mature pollen is genetically nonviable:

- thinly plated agar with a growth base of 5% raffinose for metabolism of pollen and as an osmoticum and 15% polyethylene glycol (PEG) as an osmoticum
- 50 ppm of magnesium chloride ($MgCl_2$) as a source of magnesium
- 100 ppm of boric acid (H_3BO_4) as a source of boron
- 130 ppm of potassium phosphate (KH_2PO_4) as a source of potassium
- 270 ppm of calcium chloride ($CaCl_2$) as a source of calcium
- adjust pH to 5.7 with 0.1 N sodium hydroxide (NaOH)

Other compounds are used, however sulphates are avoided (e.g. as in $MgSO_4$) and nitrates (e.g. as in $CaNO_3$) due to the risk of forming more acids that make the pH hard to control. For some pollen, this makes little difference but other species are more sensitive to changes in pH.

4. POLLEN ALLELOPATHY

4.1   Pollen Responses to Analyzed Allelochemicals

As noted in other publications, Ana Luisa Anaya's research group was first to truly elucidate the nature of pollen allelochemicals in *Zea mays* var. *chalquiñocónico*, i.e. phenylacetic acid. Phenylacetic acid appears to cause three responses in cells. One was to uncouple mitochondria oxidative phosphorylation (between coenzyme Q and cytochrome *c*), perhaps similar to parthenin inducing chlorosis. The other two responses were altering membrane permeability, theoretically preventing rehydration and harming membranes that ensure pollen of Monocotyledonae germinates and stopping mitosis.

Victoria Roshchina's group elucidated and identified the mechanisms of pollen allelochemicals. At this stage, it remains to be tested whether terpenoid allelochemicals require pollen rehydration and exudation, which would make them likely to occur in trinucleate pollen and then function by binding on plasmalemma and perhaps are then transformed in a growing heterospecific pollen tube into free radicals that affect membranes or key enzymes.

Alternatively, the allelochemicals may be similar enough to compounds involved in normal pollen germination to directly enter heterospecific cells. Though not explicitly mentioned, this presumably occurs as the pollen that is heterospecific to the allelopathic species is just beginning to germinate (a stage too small to see under a basic compound light microscope). cAMP or cGMP could be used to turn off the enzymes involved in pollen tube production in the target pollen. On a more whole-physiological basis, the effect might be altered patterns of respiration, rapid degradation of any stored carbohydrates and lipids, altered use of calcium as a chemotactic signal, or a complete depolarization of the membrane (loss of differential permeability). All of these mechanisms would stop pollen tube growth prematurely. Roshchina based these effects on the theoretical impact of terpenoids (based on their chemical structures) and it represents an impressive and useful intellectual leap, providing a better theoretical framework. The results of Roshchina's future work in this area will be enlightening.

4.2   Pollen Fluorescence in Allelopathic Response

Roshchina's group had the innovation of using microspectrofluorometry and also identified many allelochemicals. Their work micro-spectrofluorometry still stands as an excellent model that better assesses the *in vitro* and *in situ* and the mechanisms of allelopathic pollen as it measures the spectral shifts and dampening of spectra maxima from pollen before and after it alights on stigmas and is, importantly non destructive.

Principle: Measurement of the spectral shifts and dampening of spectra maxima from pollen before and after it alights on stigmas. A similar approach has been used in our experiments based on the registration of pollen fluorescence spectra.

Procedure: Using Roshchina's (1996) microspectrofluorometry technique, the author did some preliminary work in applying this to pollen of *Hieracium canadense* and its transfer to the sympatric, confamilial, and concurrently flowering *Sonchus arvensis* and *Sonchus oleraceus*. When an excitation wavelength of 435 nm is used, *Hieracium canadense* pollen may emit a yellow wavelength, i.e. 520 nm (Figure 1) while pollen from both *Sonchus* species do not.

Experiment 1. *The measurement of pollen fluorescence spectra in allelopathic studies*

The experiment followed the procedure outlined above, i.e. based on Roshchina's (1996) method and applying to the species listed for 100 runs each. The author concurs with Roshchina and Melnikova - this approach

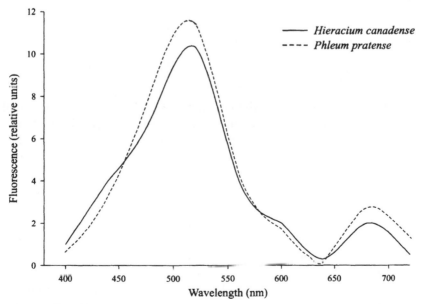

Fig. 1 The fluorescence spectrum of pollen from plant species *Hieracium canadense* and *Phleum pretense*.

may indicate exudation of flavonoids or perhaps free radicals and leads to a hypothesis that any allelopathic effects may be part of these large classes of compounds. However inconsistent results with this technique were found, i.e. emission occurred in 56 of 100 experimental runs. Therefore, a more thorough examination of the impacts of age of pollen, stigma, laboratory and ambient conditions is needed in my case. Similar results were obtained for *Phleum pretense* when its pollen alights on stigmas of *Elytrigia repens*, *Danthonia spicata*, and *Danthonia compressa* as it too emits at 520 nm (Figure 1); pollen from the other species does not emit at this wavelength.

Experment I. *Effects of pollen compounds on the germination of pollen from other species and genera.*

Principle: Use of pollen extracts and pure flavonoids from the extracts on the germination of pollen-acceptor.

Materials required: Autoclave, ion-exchange chromatograph (Murphy and Aarssen 1995a,b,c,d)

Procedure: As *Phleum pretense* produces much more pollen than *Hieracium* species and previous work, I have focused my limited efforts in this area of research on *P. pratense*. The author still uses simple ion-exchange chromatography to extract pigments and in *P. pratense*, most of these are flavonoids.

Experiment 2. *Effects of rhamnetin, isorhamnetin, and total extract from pollen-donor on the germination of pollen acceptor.*

Materials required: Autoclave, high performance liquid chromatograph, methanol, 2N HCl, pollen, Millipore UF Plus water purification system.

Procedure: Flavonoids are then further purified with 2 ml of methanolic HCl (2 N), followed by centrifugation (2 min, 15 600 g), hydrolyzation of 150 il of suspension in an autoclave (15 min, 120 C). A reverse osmosis-Millipore UF Plus water purification system is used in high performance liquid chromatography (HPLC) with an autosampler. After injections of 5 μg of samples, the mobile phases flow at a rate of 1 ml/minute with isocratic elution in a column at 30 C.

At present, it has only been possible to isolate and identify high concentrations of one compound, isorhamnetin, from pollen of *Phleum pretense* using high-speed counter-current chromatography. This does not reflect the lack of proper technology, but rather the preparation, time and expense of isolating cryptic and perhaps ephemeral compounds via techniques involving more sophisticated HPLC and mass spectrometry. Consistent with the author's earlier hypothesis that an isorhamnetin class compound is involved in pollen allelopathy in *Phleum pratense*, it is reported that if commercial rhamnetin or isorhamnetin obtained by transformation of commercial rhamnetin are added, in amounts related to their presence in pollen, to *in vitro* media or applied to stigmas, there is an allelopathic effect with isorhamnetin only (Figures 2 and 3).

The effects are significantly relative to controls but not as strong as when crude extracts are applied (Figures 2 and 3). Either other chemicals are involved in synergistic allelopathic impacts or it is a derivative or related compound to isorhamnetin that is the main allelochemical. Isorhamnetin may be the constitutive compound but stronger allelopathic effects only occur during certain phases of rehydration and then chemical exudation of pollen of *P. pratense* on a stigma. Flavonoids like isorhamnetin can cause protein kinases and specific enzymes like 6-phosphogluconate dehydrogenase that allows inositols to form pectins needed for pollen tubes.

One key issue is why *Phleum pretense* does not exhibit autotoxicity. Its pollen does fluoresce at 520 nm on conspecific stigma but it is possible that this masks more subtle changes in pollen chemistry or metabolism and detoxification of conspecific allelochemicals, *in situ*. Therefore, it is yet to be elucidated if the specific mechanism prevents autotoxicity though the

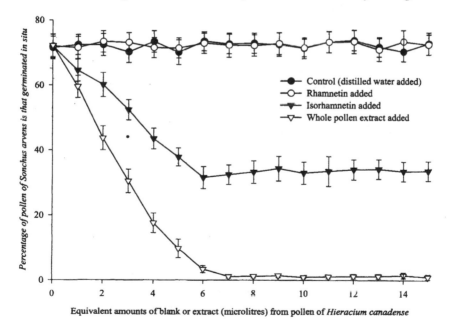

Fig. 2 Relative effects of rhamnetin, isorhamnetin and total extract from pollen of *Phleum pratense* on *in vitro* germination of pollen of *Elytrigia repens*. Data points are means and standard errors from 50 replicate samples.

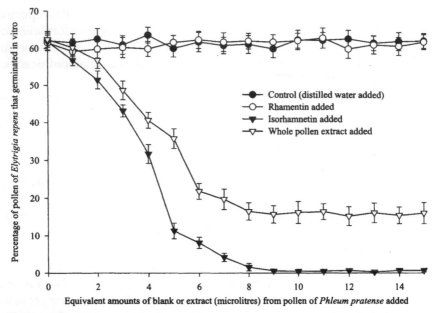

Fig. 3 Relative effects of rhamnetin, isorhamnetin and total extract from pollen of *Phleum pratense* on *in situ* (stigmatic) germination of pollen of *Elytrigia repens*. Data points are means and standard errors from 50 replicate samples.

empirical evidence of a lack of autotoxicity is important. As noted in a previous article, Roshchina's work on how terpenoid allelochemicals function may be important to a hypothesis explaining the lack of autotoxicity, if all pollen allelochemicals are able to penetrate conspecific pollen tubes and with the appropriate enzymes not found in heterospecific pollen tubes, would remove free radicals and halting any auto-allelopathic effects.

5. FUTURE RESEARCH IN POLLEN ALLELOPATHY

Given the advances in rapid assay and sophisticated chemical analyses, it is suggested that what is needed is a willingness to focus on using HPLC and MS techniques to identify and isolate allelochemicals and then introduce isolated allelochemicals into pollen cultures (and then living stigma) with the goal of examining if allelochemicals can be detected intact in con- and heterospecific pollen, if these are transformed chemically, and if enzymes in heterospecific pollen are in fact targeted, and if membrane permeability does change. At a cytoplasmic scale, such investigations are feasible but many of us (and this emphatically includes me) have focused more on the ecological implications of allelopathic pollen. It is probably time for researchers, the very authors in this volume in fact, to consider a more systematic approach and an international effort on allelopathic (pollen and other forms).

6. ACKNOWLEDGEMENTS

I thank Victoria Roshchina for the invitation to write this chapter - it has been a privilege – and Dr. S.S. Narwal for excellent suggestions and guidance. It has been a pleasure consulting with Ana Luisa Anaya, Roccio Cruz-Ortega, Stephen Gliessman, Inderjit, Azim Mallik and many others over the years. As I have noted elsewhere, Sukhada Kanchan, Jayachandra, James Thomson, Brenda Andrews, and Chris Plowright were the seminal people in early research on pollen allelopathy and we should be mindful of those who came up with the ideas carried forward by the rest of us.

7. SUGGESTED READINGS

Anaya, A. L., Hernandez-Bautista, B. E., Jimenez-Estrada, M. and Velasco-Ibarra, L. (1992). Phenylacetic acid as a phytotoxic compound of corn pollen. *Journal of Chemical Ecology* **18**: 897-905.

Anaya, A. L., Ramos, L., Hernandez, J. and Ortega, R.C. (1987). Allelopathy in Mexico. In: *Allelochemicals: Role in Agriculture and Forestry* G. R. Waller (ed), ACS Symposium Series **330**: 89-101. American Chemical Society, Washington DC.,USA

Andriotis, V. M. E. and Ross, J. D. (2003). Isolation and characterisation of phytase from dormant *Corylus avellana* seeds. *Phytochemistry* **64**: 689-699.

Brown, B. J. and Mitchell, R. J. (2001). Competition for pollination: Effects of pollen of an invasive plant on seed set of a native congener. *Oecologia* **129**: 43-49.

Brown, B. J., Mitchell, R. J. and Graham, S. A. (2002). Competition for pollination between an invasive species (purple loosestrife) and a native congener. *Ecology* **83**: 2328-2336.

Char, M. B. S. (1977). Pollen allelopathy. *Naturweissenschaften* **64**: 489-490.

Chauhan, S. V. S., Gaur, S. and Rana, A. (2005). Effect of weeds pollens on the germination of crops pollens. *Allelopathy Journal* **15**: 295-302.

Cheng, C.H. and McComb, J.A. (1992). *In vitro* germination of wheat pollen on raffinose medium. *New Phytologist* **120**: 459-462.

Cuyckens, F. and Claeys, M. (2004). Mass spectrometry in the structural analysis of flavonoids. *Journal of Mass Spectrometry* **39**: 1-15.

Fei, S. and Nelson, E. (2003). Estimation of pollen viability, shedding pattern and longevity of creeping bentgrass on artificial media. *Crop Science* **43**: 2177-2181.

Jimenez, J.J., Schultz, K., Anaya. A. L., Hernandez, J. and Espejo, O. (1983). Allelopathic potential of corn pollen. *Journal of Chemical Ecology* **9**: 1011-1025

Luczkiewcz, M., Glod, D., Baczek, T. and Bucinski, A. (2004). LC-DAD UV and LC-MS for the analysis of isoflavones and flavones from *in vitro* and *in vivo* biomass of *Genista tinctoria* L. *Chromatographia* **60**: 179-185.

Lukacin, R., Groning, I., Schiltz, E., Britsch, L. and Matern, U. (2000). Purification of recombinant flavanone 3 beta-hydroxylase from Petunia hybrida and assignment of the primary site of proteolytic degradation. *Archives of Biochemistry and Biophysics* **375**: 364-370.

Ma, C. J., Li, G. S., Zhang, D. L., Liu, K. and Fan, X. (2005). One step isolation and purification of liquiritigenin and isoliquiritigenin from *Glycyrrhiza uralensis* Risch using high-speed counter-current chromatography. *Journal of Chromatography A* **1078**: 188-192.

Martearena, M. R., Blanco, S. and Ellenrieder, G. (2003). Synthesis of alkyl-alpha-L-rhamnosides by water soluble alcohols enzymatic glycosylation. *Bioresource Technology* **90**: 297-303.

Martin, F. W. (1970). Pollen germination on foreign stigmas. *Bulletin of the Torrey Botanical Club* **97**: 1-6.

Murphy, S.D. (1992). The determination of the allelopathic potential of pollen and nectar. In: *Modern Methods of Plant Analysis. Volume 13. Plant Toxin Analysis*, H.F. Linskens and J.F. Jackson (eds), pp. 333-357. Springer-Verlag, New York,USA

Murphy, S. D. (1999). Is there a role for pollen allelopathy in biological control of weeds? In: *International Allelopathy Update - Volume 2. Basic and Applied Aspects*, S. S. Narwal (ed). pp. 204-215. Science Publishers, Enfield, NH,USA.

Murphy, S. D. (1999). Pollen allelopathy. In: *Principles and Practices in Plant Ecology*, Inderjit, K. M. Dakshini and C. L. Foy (eds), pp. 129-148 CRC Press, Boca Raton, Florida, USA

Murphy, S. D. (2001). Field testing for pollen allelopathy: A review. *Journal of Chemical Ecology* **26**: 2155-2172

Murphy, S. D. (2001). The role of pollen allelopathy in weed ecology. *Weed Technology* **15**: 867-872.

Murphy, S. D. and Aarssen, L. W. (1989). Pollen allelopathy among sympatric grassland species: *In vitro* evidence in *Phleum pratense* L. *New Phytologist* **112**: 295-305.

Murphy, S. D. and Aarssen, L. W. (1995). *In vitro* allelopathic effects of pollen from three *Hieracium* species (Asteraceae) and pollen transfer to sympatric Fabaceae. *American Journal of Botany* **82**: 37-45.

Murphy, S. D. and Aarssen, L. W. (1995). Allelopathic pollen extract from *Phleum pratense* L (Poaceae) reduces germination, *in vitro*, of pollen in sympatric species. *International Journal of Plant Science* **156**: 425-434.

Murphy, S. D. and Aarssen, L. W. (1995). Allelopathic pollen extract from *Phleum pratense* L (Poaceae) reduces seed set in sympatric species. *International Journal of Plant Science* **156**: 435-444.

Murphy, S. D. and Aarssen, L. W. (1995). Allelopathic pollen of *Phleum pratense* reduces seed set in *Elytrigia repens* in the field. *Canadian Journal of Botany* **73**: 1417-1422.

Murphy, S. D. and Aarssen, L. W. (1996). Partial cleistogamy limits reduction in seed set in *Danthonia compressa* (Poaceae) by allelopathic pollen of *Phleum pratense* (Poaceae). *Écoscience* **3**: 205-210.

Nyiredy, S. (2004) Separation strategies of plant constituents-current status. *Journal of Chromatography B-Analytical Technologies in the Biomedical and Life Sciences* **812**: 35-51.

Ollila, F., Halling, K., Vuorela, P., Vuorela, H. and Slotte, J. P. (2002). Characterization of flavonoid-biomembrane interactions. *Archives of Biochemistry and Biophysics* **399**: 103-108.

Ormerod, M. (2000). *Flow Cytometry: A Practical Approach*. Third edition. Oxford University Press, Toronto, Canada

Ortega, R.C., Anaya, A.L. and Ramos, L. (1988). Effects of allelopathic compounds of corn pollen on respiration and cell division of watermelon. *Journal of Chemical Ecology* **14**: 71-86.

Parry, A. D. and Edwards, R. (1994). Characterization of o-glucosyltransferases with activities toward phenolic substrates in alfalfa. *Phytochemistry* **37**: 655-661.

Phamdelegue, M. H., Loublier, Y., Ducruet, V., Douault, P., Marilleau, R. and Etievant, P. (1994). Characterization of chemicals involved in honeybee-plant interactions. *Grana* **33**: 184-190.

Prigione, V., Lingua, G. and Marchisio, V. F. (2004). Development and use of flow cytometry for detection of airborne fungi. *Applied and Environmental Microbiology* **70**: 1360-1365.

Radlowski, M., Kalinowski, A., Adamczyk, J., Krolikowski, Z. and Bartkowiak, S. (1996). Proteolytic activity in the maize pollen wall. *Physiologia Plantarum* **98**: 172-178.

Roshchina, V.V. (1999). Mechanisms of cell-cell communication. In: *International Allelopathy Update - Volume 2. Basic and Applied Aspects* S. S. Narwal (ed). pp. 3-26. Science Publishers, Enfield, NH, USA.

Roshchina, V. V. (2001). Molecular-cellular mechanisms in pollen allelopathy. *Allelopathy Journal* **8**: 11-28.

Roshchina, V. V. (2004). Cellular models to study the allelopathic mechanisms. *Allelopathy Journal* **13**: 3-15.

Roshchina, V. V. (2005). Allelochemicals as fluorescent markers, dyes and probes. *AllelopathyJournal* **16**: 31-46.

Roshchina, V. V. and Melnikova, E. V. (1996). Microspectrofluorometry: A new technique to study pollen allelopathy. *Allelopathy Journal* **3**: 51-58

Roshchina, V. V. and Melnikova, E. V. (1998). Allelopathy and plant generative cells: Participation of acetylcholine and histamine in signaling interactions of pollen and pistil. *Allelopathy Journal* **5**: 171-182

Roshchina, V.V. and Melnikova, E. V. (1999). Microspectrofluorometry of intact secreting cells applied to allelopathy. In: *Principles and Practices in Plant Ecology* ., Inderjit, K. M. M. Dakshini, and C. L. Foy (eds) pp. 99-126. CRC Press, Boca Raton, Florida, USA.

Shapiro, H. M. (2003). *Practical Flow Cytometery*. Wiley-Liss, New YorkUSA

Sukhada, K.D. and Jayachandra (1980a). Pollen allelopathy - a new phenomenon. *New Phytologist* **84**: 739-746

Sukhada, K.D. and Jayachandra (1980b). Allelopathic effects of *Parthenium hysterophorus* L. Part IV. Identification of inhibitors. *Plant and Soil* **55**: 67-75

Thomson, J.D., Andrews, B.J. and Plowright, R.C. (1982). The effect of a foreign pollen on ovule development in *Diervilla lonicera* (Caprifoliaceae). *New Phytologist* **90**: 777-783

Vanrensen, I. and Veit, M. (1995). Simultaneous determination of phenolics and alkaloids using ion-exchange chromatography for sample preparation. *Phytochemical Analysis* **6**: 121-124.

Viswanathan, K. and Lakshmanan, K.K. (1984). Phytoallelopathic effects on in vitro pollinnial germination of *Calotropis gigantea* R. Br. *Indian Journal of Experimental Botany* **22**: 544-547.

Wolfender, J. L., Waridel, P., Ndjoko, K., Hobby, K. R., Major, H. J. and Hostettmann, K. (2000). Evaluation of Q-TOF-MS/MS and multiple stage IT-MSn for the dereplication of flavonoids and related compounds in crude plant extracts. *Analysis* **28**: 895-906A.

Xu, P, Vogt, T. and Taylor, L.P. (1997). Uptake and metabolism of flavonols during in-vitro germination of *Petunia hybrida* (L) pollen. *Planta* **202**: 257-265.

Subject Index

T - #0132 - 111024 - C224 - 229/152/11 - PB - 9780367446222 - Gloss Lamination